T0230903

Ballast Railroad Design: SMART-UOW Approach

Ballast Railroad Design: SMART-UOW Approach

Buddhima Indraratna

Centre for Geomechanics and Railway Engineering,
University of Wollongong, Australia

Trung Ngo

Centre for Geomechanics and Railway Engineering,
University of Wollongong, Australia

CRC Press
Taylor & Francis Group
Boca Raton   London   New York   Leiden

CRC Press is an imprint of the
Taylor & Francis Group, an **informa** business

A BALKEMA BOOK

Cover illustrations: Courtesy of Dr Fernanda Ferreira (Research Associate),
Alan Grant (Technical Staff) and Dr Navaratnarajah (former PhD student)

CRC Press/Balkema is an imprint of the Taylor & Francis Group, an informa
business

© 2018 Taylor & Francis Group, London, UK

Typeset by Apex CoVantage, LLC

Library of Congress Cataloging-in-Publication Data
Names: Indraratna, Buddhima, author. | Ngo, Trung, author.
Title: Ballast railroad design : SMART-UOW approach / Buddhima Indraratna,
    Centre for Geomechanics and Railway Engineering, University of
    Wollongong, NSW 2522, Australia, Trung Ngo, Centre for Geomechanics
    and Railway Engineering, University of Wollongong, NSW 2522, Australia.
Description: First edition. | Boca Raton : CRC Press/Balkema, [2018] |
    Includes bibliographical references and index.
Identifiers: LCCN 2018011009 (print) | LCCN 2018014998 (ebook) | ISBN
    9780429504242 (ebook) | ISBN 9781138587038 (hardcover : alk. paper)
Subjects: LCSH: Ballast (Railroads)
Classification: LCC TF250 (ebook) | LCC TF250 .I535 2018 (print) | DDC
    625.1/41—dc23
LC record available at https://lccn.loc.gov/2018011009

Published by: CRC Press/Balkema
Schipholweg 107c, 2316 XC Leiden, The Netherlands
e-mail: Pub.NL@taylorandfrancis.com
www.crcpress.com – www.taylorandfrancis.com

ISBN: 978-1-138-58703-8 (Hbk)
ISBN: 978-0-429-50424-2 (eBook)

Contents

Preface

Railway track systems are constructed to provide a smooth and safe transport mode for passengers or freight trains. They are designed to sustain the stresses imposed by lateral, longitudinal and vertical loads acting on the track structure. A ballasted railway track system comprises several components, among which steel rails, rail fasteners, timber, steel or concrete sleepers, granular ballast, sub-ballast and subgrade materials are the main constituents. The recent increases in axle loads, speed and traffic volume, along with the need to improve passenger comfort and reduce track life cycle costs, have created a need for track design optimisation. Furthermore, complementary decision support systems require a more precise analytical and mechanistic approach to meet the design needs of modern railway track systems. These aspects highlight the necessity of a thorough review and revision of the current railway track design.

Given the lack of capacity of current ballasted tracks in many parts of Australia to support increasingly heavier and faster trains, the development of innovative and sustainable ballasted tracks is crucial for transport infrastructure. Ballast degradation and infiltration of fine particles such as coal along the heavy haul corridors and soft subgrade soils contaminating the overlying ballast decrease the porosity of the ballast layer and impede track drainage. This leads to excessive track settlements and instability, as well as increased maintenance costs. To mitigate these problems, the utilisation of geosynthetics (e.g. polymer geogrids, geocomposite, geocells) and recycled rubber mats has been investigated by the authors.

The tangible outcomes of this research study has made a considerable impact on industry in view of forcing design modifications and provision of new technical standards for Australian railways. Already, a considerable portion of the R&D work in this area of research is captured in our in-house computer software (SMART – supplementary methods of analysis for railway track), which can accommodate a variety of problematic ground conditions in Australia in user-friendly modules that enable best track management practices.

This book presents a comprehensive procedure of ballasted track design based on a rational approach that combines extensive laboratory testing, mathematical and computational modelling and field measurements carried out over the past two decades. The Ballast Railroad Design: SMART-UOW approach can be regarded as a useful guide to assist the practitioner, rather than a complete design tool to replace existing rational design approaches. Practising engineers can refer to this book for designing new tracks as well as to remediate existing ballasted tracks with subgrade deformation problems because it provides a systematic approach and greater flexibility in track design. This book can also be used as a useful resource by postgraduate students and as a teaching tool by academics in track design and maintenance.

Buddhima Indraratna
Trung Ngo

Foreword

Studies on ballasted rail tracks have been conducted at the University of Wollongong for more than two decades, and these research outcomes have significantly influenced the way that rail tracks can be modernised through innovative design. Imparting that knowledge to today's rail practitioners, especially those in heavy haul operations, is the objective of Ballast Railroad Design: SMART-UOW approach. This book complements the software SMART (*s*upplementary *m*ethods of *a*nalysis for *r*ailway *t*rack) currently managed by the University of Wollongong together with the Australasian Centre for Rail Innovation (ACRI).

This book deals with both theoretical and practical issues directly related to ballasted tracks, considering a series of options from the selection of mechanical and geotechnical parameters to advanced design examples, capturing the influence of various factors such as particle breakage, ballast fouling, track confining pressure and the application of geosynthetics. The technical content also assists in track maintenance incorporating subgrade deformation and stability considerations, supplemented by case studies and large-scale simulations. Importantly, complex technical content is presented for practitioners in a clear and concise manner, working through examples based on real world situations.

With significantly increased axle loads and speeds of freight trains supporting the mining and agriculture industries in many nations, including Australia, design and construction requirements, and longevity and performance expectations, have become increasingly strategic and challenging than the traditional heavy haul tracks of the past. This is a timely book presenting considerations for contemporary track design and current state-of-the-art practice in ballast railroads. It has been informed through collaborative research with industry, incorporating sophisticated laboratory tests, computational modelling and field studies to advance the design of ballasted tracks.

ACRI congratulates the University of Wollongong on this enhancement to SMART and associated railroad design and analysis and the contribution it will make to the rail industry through informing engineering solutions and advancing industry training.

Andrew Meier,
CEO, Australasian Centre for Rail Innovation

About the authors

Distinguished Professor Buddhima Indraratna (FTSE, FIEAust, FASCE, FGS, FAusIMM) is a civil engineering graduate from Imperial College, London, and obtained his PhD from the University of Alberta in 1987. He worked in industry in several countries before becoming an academic and has been a United Nations Expert and Foreign Advisor to numerous overseas projects. Professor Indraratna's pioneering contributions to railway geotechnology and various aspects of geotechnical engineering have been acknowledged through numerous national and international awards, including the 1st Ralph Proctor Lecture and 4th Louis Menard Lecture of the International Society of Soil Mechanics and Geotechnical Engineering, ISSMGE; the 2015 Thomas Telford Premium Award (ICE, UK); the 2009 EH Davis Memorial Lecture of Australian Geomechanics Society; and the 2014 CS Desai Medal for his substantial and sustained contributions to Transport Geotechnics and Ground Improvement, respectively. Recently, he was the recipient of the 2017 Outstanding Contributions Medal of IACMAG. The New South Wales Minister of Transport awarded Professor Indraratna the 2015 Australasian Railway Society's Outstanding Individual Award at the State Parliament. His pioneering contributions to railway engineering and ground improvement earned him the Fellowship of the Australian Academy of Technological Sciences and Engineering (FTSE) in 2011.

Distinguished Professor Indraratna has made fundamental contributions to transport geotechnology and ground improvement. He has developed unique process simulation equipment for geomaterials and computational methods for predicting the dynamic response of transport infrastructure. His original efforts capture the role of particle breakage in heavy haul rail environments, the only theory to determine the optimum confining pressure for railroad stability, the methods of eliminating impact-based track damage using energy-absorbing recycled materials, innovative computer simulations of vacuum pressure application for stabilising foundations and field-based methods for quantifying soil disturbance. His research has influenced national and international standards and projects in road/rail embankments and port reclamation.

Dr Trung Ngo is an early career researcher with internationally recognised expertise in the field of physical and computational modelling of ballasted rail tracks using Discrete Element Method (DEM) and Finite Element Method (FEM). After graduating from the Ho Chi Minh city University of Technology, Viet Nam, he obtained a master's and a PhD from the University of Wollongong, Australia, under the supervision of Distinguished Professor Indraratna. In 2013, Dr Ngo's doctoral research was acknowledged by Railway Technical Society of Australasia (RTSA) and he was honoured by RTSA Postgraduate Thesis Award, which is awarded once every three years, recognising "contributions to rail industry in transferring the results of advanced computational, theoretical, and laboratory research into professional engineering practices". Dr Trung Ngo is currently a research fellow at the Centre for Geomechanics & Railway Engineering (CGRE), University of Wollongong, working under the Rail Manufacturing CRC project. His research has primarily focused on the area of railway track design, specifically in computational modelling and laboratory testing for ballasted rail tracks.

Acknowledgements

This book was introduced on the basis of knowledge acquired through two decades of laboratory studies, field observations and computational studies on railroad engineering conducted at the Centre for Geomechanics and Railway Engineering (CGRE), University of Wollongong, Australia. It contains research deliverables of numerous sponsored projects completed since the mid-1990s. Many of the concepts and analytical principles incorporated herein have already been described to some extent in the more elaborated textbook *Advanced Rail Geotechnology – Ballasted Track* and in various peer-reviewed research papers published by the first author and his co-workers and research students.

Significant contributions made over the years by Dr Sanjay Nimbalkar, A/Prof Cholachat Rujikiatkamjorn and A/Prof Hadi Khabbaz through their involvement in numerous sponsored ARC and CRC projects are gratefully acknowledged. The authors specifically thank industry colleagues, Tim Neville (ARTC) and, more recently, Dr Richard Kelly (SMEC), for pointing out the imperative need for such a publication to assist modern track designers.

The authors also would like to acknowledge the well-known Australian senior rail practitioners David Christie (formerly at RIC and RailCorp), Mike Martin (formerly at Queensland Rail) and Jatinder Singh (Sydney Trains) for their active collaboration over many years. Particular mentions with gratitude go to Prof ET (Ted) Brown and Prof Harry Poulos, who have supported and inspired the first author over many years, encouraging him to pioneer cutting-edge research in track geotechnology, as this field of research was lacking in Australia. The support received through the ARC Centre of Excellence for Geotechnical Science and Engineering (CGSE) together with three consecutive rail-based Cooperative Research Centres (CRC) during the past decade are gratefully acknowledged.

The contents captured in this book are attributed to the original efforts of many research students and staff at the Centre GRE, University of Wollongong. Contributions at various times by A/Prof Jayan Vinod, Dr Ana Heitor, Dr Jahanzaib Israr, Dr Qideng Sun and Dr Fernanda Ferreira are acknowledged. The specific research works of former PhD students Dr Wadud Salim, Dr Joanne Lackenby, Dr Nayoma Tennakoon, Dr Khaja Karim Hussaini, Dr Yifei Sun, Dr Daniel Ionescu, Dr Dominic Trani, Dr Sinnaiah Navaratnarajah, Dr Mahdi Biabani and Dr Pramod Thakur, among others, are also captured in the content in various forms. The authors are also grateful to UOW technical staff Alan Grant, Cameron Neilson, Duncan Best, Frank Crabtree and Ritchie McLean for their assistance in laboratory and field work.

A number of important research projects on ballasted rail tracks and geosynthetics have been supported in the past and are currently supported by the Australian Research Council (ARC) through its Discovery and Linkage programs. Keen collaboration with industry partners has facilitated the application of theory into practice. In this respect, the authors greatly appreciate the financial support from the Rail Manufacturing Cooperative Research Centre (funded jointly by participating rail organisations and the Australian federal government's Business Cooperative Research Centres Program). The authors also appreciate the Australasian Centre for Rail Innovation (ACRI), Tyre Stewardship Australia (TSA), Global Synthetics Pty Ltd, Foundation Specialists Group, Sydney Trains and ARTC, among others.

Selected technical data presented in numerous figures, tables and some technical discussions have been reproduced with the kind permission of various publishers, including: *Géotechnique*, *ASCE Journal of Geotechnical and Geoenvironmental Engineering*, *Computers and Geotechnics*, *International Journal of Geomechanics*, *Geotextiles and Geomembranes* and *Geotechnical Testing Journal*, among others; salient content from these previous studies are reproduced here with kind permission from the original sources.

Finally, the authors also acknowledge the efforts of Dr Udeni Nawagamuwa, Mrs Manori Indraratna and Mr Bill Clayton for their assistance during copy editing and proofreading of the contents.

Chapter 1

Introduction

1.1 General background

Rail networks form an important part of the transport system in Australia and many other countries in the world. Railways play a vital role in its economy by transporting freight and bulk commodities between major cities and ports and by carrying passengers, particularly in urban areas. The Australian rail has carried around one-third of all domestic freight over the past 25 years, and millions of passengers travel in trains each year. For instance, the longest and heaviest train in Western Australia has had a gross weight of nearly 100,000 tonnes and a length exceeding 7 km, with as many as 682 wagons hauled by eight locomotives (Railway Gazette 2001). The need to maintain a competitive edge over other means of transportation has increased the pressure on the railway industry to improve its efficiency and decrease maintenance and infrastructure costs (Indraratna *et al.* 2011a). With ballasted railway tracks, the cost of substructure maintenance can be significantly reduced with a better understanding of the physical and mechanical characteristics of the rail substructure and the ballast layer in particular.

In a ballasted rail track, a large portion of the track maintenance budget is spent on ballast-related problems (Indraratna *et al.* 2011b). Although ballast usually consists of hard and strong angular particles derived from high strength un-weathered rocks, it also undergoes gradual and continuing degradation under cyclic rail loadings (Indraratna *et al.* 2011a; Selig and Waters 1994). The sharp edges and corners are broken due to high stress concentrations at the contact points between adjacent particles. The reduction in angularity decreases its angle of internal friction (i.e. shear strength), which in turn increases plastic settlement of the track.

In low-lying coastal areas where the subgrade soils are generally saturated, the fines (clays and silt-size particles) can be pumped up into the ballast layer as slurry under cyclic rail loading, if a proper subbase or filter layer is absent (Raut 2006; Selig and Waters 1994). The pumping of subgrade clay is a major cause of ballast fouling. Fine particles from clay pumping or ballast degradation form a thin layer surrounding the larger grains that increases compressibility, fills the void spaces between larger aggregates, and reduces the drainage characteristics of the ballast bed (Indraratna *et al.* 2014). The fouling of ballast usually increases track settlement and may cause differential track settlement. Where there is satu-ration and poor drainage, any contamination of ballast may also cause localised undrained failure. In severe cases, fouled ballast needs to be cleaned or replaced to keep the track up to its desired stiffness (resiliency), bearing capacity, alignment and level of safety (Indraratna *et al.* 2013a; Tennakoon *et al.* 2012).

1.2 Limitations of current track design practices

The most common track design approach nowadays follows the Li and Selig (1998a, b) method, but this method has several drawbacks because it does not consider: (i) the effect of ballast breakage, cyclic loading, confining pressure and multiple subgrade layers; (ii) permeability reduction due to fine contamination; (iii) fouling assessment and implications on speed restrictions; and (iv) the use of geosynthetics in tracks.

1.3 New developments in SMART-UOW approach

The conventional track design methods commonly assume the track foundation is an elastic media using the Winkler model (i.e. a continuously supported beam on an elastic foundation) currently used in track design practices. This approach has several unrealistic assumptions, including static point loads for the trains; continuous support under the rail; linear characteristics for the track support system; and linear uniform pressure distribution under the sleeper, among others.

Although several factors have been taken into account to compensate for the errors caused by these assumptions, there can be significant discrepancies between the results obtained from conventional methods and those obtained in railway fields. This indicates that the effectiveness of conventional track design is questionable due to two aspects: cost efficiency and accuracy. In the past, some improvements have been made to provide a better understanding of railway track systems; these new developments can be used to improve the current track design approach.

This book presents a supplementary methodology for ballasted railway track design and maintenance supported by extensive laboratory tests and field measurements carried out by the Center for Geomechanics and Railway Engineering, University of Wollongong (UOW), over the past two decades.

The main Ballast Railroad Design: SMART-UOW approach introduced herein includes several modules that allow railway practitioners to input required design parameters and perform track design, including:

- Module 1: Input design parameters for track design
- Module 2: Check the allowable bearing capacity of the ballasted tracks
- Module 3: Design thickness of the granular layer (ballast, capping layer, structural fill)
- Module 4: SMART-UOW track design approach that allows designers to:

  - Predict ballast breakage
  - Consider the effect of the confining pressure and loading frequency on track performance
  - Quantify ballast fouling and associated track drainage and train speed
  - Application of geosynthetics in tracks
  - Predict vertical track settlement
  - Incorporate constitutive model for ballast
  - Design of capping layer (sub-ballast) with filtration properties

UOW design approach is presented in a flowchart in Figure 1.1.

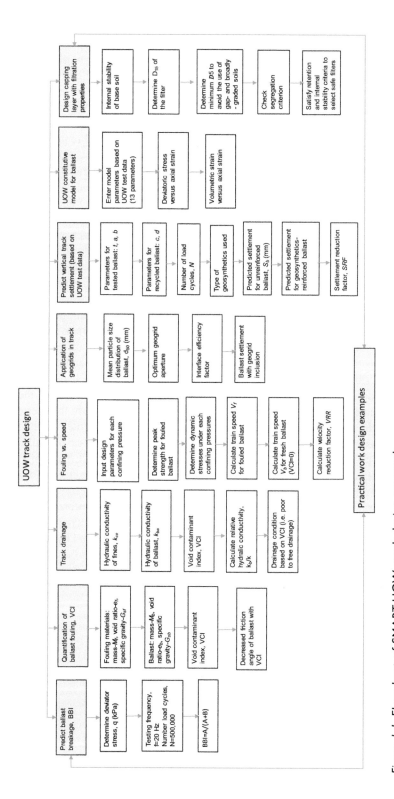

Figure 1.1 Flowchart of SMART-UOW track design approach

1.4 Scope

This book presents creative and innovative solutions to rail industry worldwide and is the result of knowledge acquired through two decades of laboratory studies, field observations and computational studies on railroad engineering conducted at the Centre for Geomechanics and Railway Engineering (CGRE), University of Wollongong, Australia. Keeping the critical issues of track substructure in mind, the authors present the current state of research, concentrating on: (i) the procedure to determine bearing capacity and required thickness of the granular layer of ballasted tracks; to consider the effect of the confining pressure and loading frequency on track performance; (ii) the effectiveness of various geosynthetics in minimising ballast breakage and controlling track settlement; (iii) the role of constitutive modelling of ballast under cyclic loading, the effect of ballast fouling and its implications on track performance; (iv) the design of sub-ballast and the filtration layer; and (v) practical worked-out design examples. The authors hope that this book will not only become an imperative design aid for practitioners but will also be a valuable resource for postgraduate students and researchers in railway engineering. The book will generate further interest among both researchers and practicing engineers in the wide field of rail track geotechnology and promote much needed track design modifications.

Chapter 1 describes the general background of rail networks, limitations of current track design practices and the new developments in SMART-UOW track design approach. Chapter 2 presents key parameters needed for ballasted track design, different components of track substructure, typical ballasted track problems, design criteria and traffic conditions. Chapter 3 describes a procedure to determine the bearing capacity of ballasted tracks. The details of the design procedure and a flowchart to determine the thickness of the granular layer are presented in Chapter 4. Chapter 4 also describes a procedure to determine track modulus and the resilient modulus of ballast. Chapter 5 presents studies on the effects of confining pressure and frequency on ballast breakage. The influence of ballast fouling and implications for track performance are discussed in Chapter 6. Effects of fouling on the drainage capacity of track and operational train speed are also discussed in Chapter 6. Chapter 7 presents the use of geosynthetics in railway tracks and the effects of coal fouling on the inter-particle friction angle and load-deformation of geogrid-reinforced ballast. A new stress–strain constitutive model for ballast incorporating particle breakage is discussed in Chapter 8. Chapter 9 describes a design procedure for sub-ballast and the filtration layer. Practical design examples are presented in Chapter 10. Finally, the introduction of SMART software (supplementary methods of analysis for railway track) to aid in the analysis and design of rail track substructure is presented.

Chapter 2

Parameters for track design

2.1 General background

The rail track network forms an important part of the transportation infrastructure in Australia. It plays a significant role in maintaining a healthy economy by transporting export-oriented heavy bulk freight (coal, minerals and agricultural products) and carrying passengers between major cities and from various inland terminals to ports. With increasing competition from other means of transport such as trucks, buses, aircraft and ships, the railway industry must continually upgrade the track system and apply innovative technologies to minimise the cost of construction and maintenance, as well as increase passenger comfort. Salim (2004) showed that Australia has more than 43,000 km of narrow, broad, standard and dual gauge ballasted rail tracks (Fig. 2.1). In Australia, millions of passengers travel in trains every year, particularly in the state of New South Wales (NSW). According to reported data (RailCorp 2010–2011), around 300 million passengers travelled by train during the financial year 2010–2011, and growth in CityRail passenger journeys increased by 1.8% in this year, leading to an approximate increase of 10% compared to 2005.

Costs associated with track maintenance are steadily increasing due to the utilisation of heavier and faster trains, as well as a lack of effective methods for strengthening the track substructure. In many nations, significant funds are invested in track maintenance to sustain stability and improve passenger comfort. The Australian government has just invested AU$ 1.3 billion in the rail industry (RailCorp, 2010–2011) to make continual improvements to the rail networks. The funds spent to maintain the track substructure, including ballast, sub-ballast and subgrade, are significant compared to those spent on the track superstructure (rails, fasteners and sleepers) (Selig and Waters 1994). The American railway industry has invested millions of dollars per year for ballast replacement and associated maintenance costs (Chrimer 1985). During the 1992–1993 period, the State Rail Authority of New South Wales used in excess of 1.3 million tonnes of ballast at a cost of over AU$ 12 million to replace the ballast in rail track (Indraratna *et al.* 1997), and millions of dollars were invested annually to quarry and purchase 800,000 tonne of coarse aggregates for ballast in New South Wales (Lackenby 2006).

2.2 Typical ballasted track problems

The ballast layer plays a crucial part in transmitting and distributing the wheel load from sleepers to the underlying sub-ballast and subgrade at a reduced and acceptable stress level (Selig and Waters 1994). It normally consists of strong, medium to coarse-sized granular

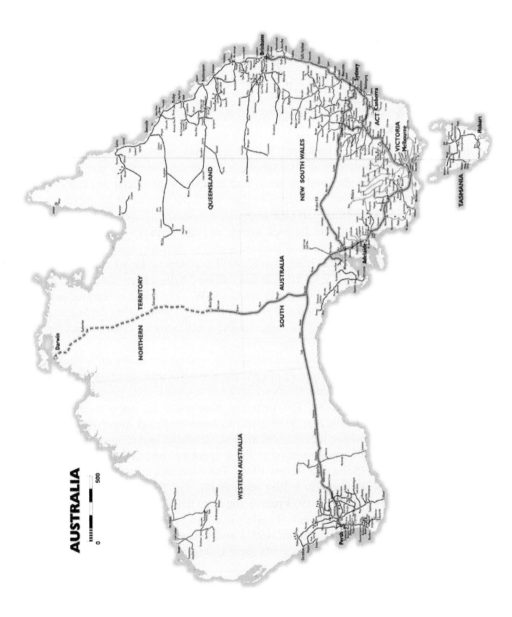

Figure 2.1 Australia's railway network
(courtesy of Australian Railroad Group)

particles (10–63 mm) with a large amount of pore space and a permeable structure to assist in rapid drainage, and it also has a high load bearing capacity (Indraratna *et al.* 2011b). During operation, ballast deteriorates due to the breakage of angular corners and sharp edges, the infiltration of fines from the surface and mud pumping from the subgrade under train loading. As a result of these actions, ballast becomes less angular, fouled, reduced in shear strength and with impeded drainage (Fig. 2.2). Fouling materials have traditionally been considered as unfavourable to track structure because they increase deformation and may cause differential track settlement. Where there is saturation and poor drainage, trapped water results in increased pore water pressure and subsequent localised undrained shear failure of the ballast.

Modernising the national railroad infrastructure is a challenge facing all developed societies due to increased competition from other means of transportation. Consequently, adopting innovative and effective methods to improve serviceability and effectiveness and reduce maintenance and infrastructure costs of rail tracks is inevitable. Walls and Galbreath (1987) showed that the periods between maintenance cycles could be increased by as much as 12 times by using geogrids to reinforce ballast. Geogrid is a type of polymer geosynthetic usually placed between the layer of sub-ballast and ballast to provide additional confining pressure and strengthen the ballast due to interlocking with surrounding ballast aggregates. As a result, this significantly decreases lateral spreading, a major cause of ballast deformation. Although the effect of geogrid in strengthening the ballast layer has been recognised, the interface behaviour between the geogrid and ballast has not been examined in detail or incorporated into ballasted track design. This is probably because when ballast is fouled, the effectiveness of geogrid is believed to decrease significantly due to fine particles clogging the apertures of the geogrid and acting as lubricant, leading to reduced interlocking and mobilised frictional resistance between the geogrid and ballast (Indraratna *et al.* 2011a; Ngo *et al.* 2017a). Therefore, the degree of improvement in track performance with the inclusion of geogrid while considering the various fouling conditions must be investigated and incorporated into existing track designs.

Figure 2.2 Section of track fouled ballast and poor drainage
(courtesy of Australian Transport Safety Bureau, 2012)

2.3 Typical input parameters for track design

The main input parameters considered in track design include: dynamic wheel loads, tonnage by million gross tonnes, moduli of granular materials and subgrade soil, subgrade soil type and compressive strength, and other necessary parameters as presented in the input parameter chart presented in Figure 2.3.

Input parameters such as width of sleeper (B), length of sleeper (L), unit weight of ballast (γ) and friction angle of ballast (ϕ) are used to determine the ultimate bearing capacity of the ballast. The factor of safety (FOS) input box is then used to calculate the allowable bearing capacity. Other input parameters, such as static wheel load (P_s), wheel diameter (D), sleeper spacing (a) and train velocity (V), are used to determine the static and dynamic stresses on the ballast sleeper interface using the AREA (1974) method.

2.4 Substructure of ballasted tracks

The components of typical ballasted rail track are generally divided into two main categories: (i) superstructure (rails, fastening system, sleepers) and (ii) substructure (ballast, sub-ballast, subgrade), as depicted in Figure 2.4. Upon repeated train loading, wheel loads are

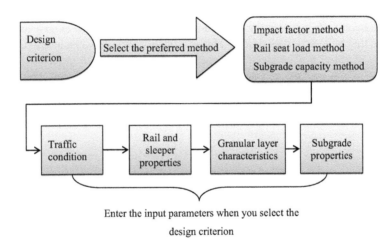

Figure 2.3 Components of required input parameters for track design

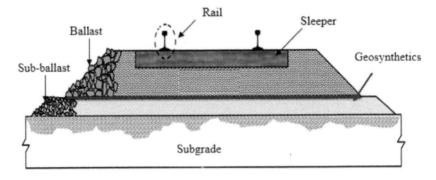

Figure 2.4 Schematic of main components of track structures

transferred from the superstructure into the substructure through the ballast layer. To deliver safety and passenger comfort, each component of the track structure must perform its desired functions properly while responding to the anticipated train loadings and environmental conditions imposed on the tracks.

2.5 Ballast

Ballast is a free draining granular material that helps transmit and distribute an induced cyclic load to the underlying sub-ballast and subgrade at a reduced and acceptable level of stress (Indraratna *et al.* 1998; McDowell *et al.* 2008). It is a natural or crushed granular material with a typical thickness of 250–450 mm that is placed beneath the track superstructure and above the sub-ballast (capping) or subgrade (Ionescu 2004; Sun *et al.* 2016). Conventionally, coarse-sized, angular, crushed, hard stones and rocks that are uniformly graded, free of dust and not prone to cementing action have been considered as good ballast materials (Lackenby *et al.* 2007; Ngo *et al.* 2014; Selig and Waters 1994; Sun *et al.* 2014a; Tutumluer *et al.* 2007). Owing to limited universal agreement on the engineering characteristics of ballast, the selection of ballast sources generally depends on availability and economic considerations. Ballast gradation conforms to the gradation limit specified in Australia (AS-22758.7 1996). Table 2.1 shows the grain size characteristics of ballast materials used by the author in the laboratory tests. The typical particle size distribution curves (PSD) of ballast used in this study are plotted in Figure 2.5, together with the PSD of sub-ballast and coal fines.

2.5.1 Ballast characteristics

Parameters characterising the mechanical properties of ballast can be entered into the program in the "input design parameter" section. The details of these parameters are provided in the following.

*   "Ballast: density" is the bulk density of ballast in the unit of t m³ or kN m³. The density should be determined from specimens with a level of compaction similar to the ballast in the actual ballasted tracks. Typical densities of ballast made of volcanic rock compacted to meet the construction standards adopted in Australia vary between 1.5 and 1.65 t m³.

Table 2.1 Grain size characteristics of ballast tested in the laboratory

Test type	Particle shape	d_{max} (mm)	d_{10} (mm)	d_{30} (mm)	d_{50} (mm)	d_{60} (mm)	C_u	C_c	Size ratio
Tested ballast gradation	Highly angular	53	16	28	35	39	2.4	1.3	6

Note:
d_{max}: maximum ballast size used in this study
d_{10}: diameter in millimetres at which 10% by weight of ballast passes through the sieve
d_{30}, d_{50}, d_{60}: diameters in millimetres at which 30%, 50% and 60% by weight of ballast passes through the sieve

C_u: coefficient of uniformity, determined by: $C_c = \dfrac{d_{60}}{d_{10}}$

C_c: coefficient of curvature, determined by: $C_c = (d_{30})^2/d_{10}/d_{60}$; $C_c = \dfrac{(d_{30})^2}{d_{10} \times d_{60}}$

Size ratio: ratio of apparatus dimension (max length) divided by maximum size of ballast

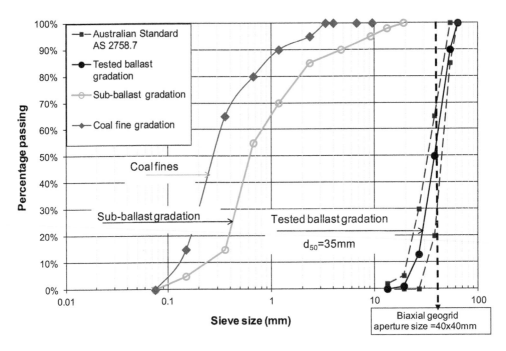

Figure 2.5 Particle size distribution of ballast, sub-ballast and coal fines used in the laboratory

- "Ballast: Poisson's ratio" is the Poisson's ratio of the ballast layer. It is the ratio of the corresponding lateral deformation to the vertical deformation of the ballast mass. Typical Poisson's ratios of ballast made of rock fragments of volcanic origin compacted to meet the construction standards adopted in Australia are between 0.3 and 0.35. It should be noted that the value of Poisson's ratio cannot theoretically be smaller than 0.1 or greater than 0.5.
- "Ballast: resilient modulus" is the resilient modulus of ballast measured from the cyclic standard or cubical triaxial tests. The modulus is taken as the slope of the deviator stress-vertical strain (q-ε_1) curve during unload and reload cycles. The magnitude of resilient modulus varies depending on the void ratio and effective confining stress to the ballast. The void ratio of ballast generally decreases with the number of load cycles due to rearrangement and breakage of the ballast particles, which results in denser packing of the ballast mass. As a result, the resilient modulus increases.
- Indraratna and Salim (2003) conducted a series of cubical triaxial tests on ballast specimens made of latite basalt and reported a decrease in the void ratio of about 0.04 when the number of load cycles was increased to 500,000. This corresponded to an increase of cyclic resilient modulus from about 60 to 200 MPa. The values of effective confining stresses used in these tests were in the range of 7–20 kPa, which are typical of the pressures generally present in conventional tracks. For design purposes, the value of resilient modulus may be approximated from the equation below using the average number of load cycles (N) expected for the design period:

$$Resilient\ modulus\ (MPa) = 60 + \frac{N}{500,000} \times 140; \quad N < 500,000 \tag{2.1}$$

- "Ballast: friction angle" (φ) is the effective friction angle of ballast in degrees. The friction angle defines the strength of the ballast and can be obtained from laboratory strength tests such as standard triaxial and direct shear tests. The value of the friction angle depends on the angularity of ballast particles and type of ballast rock, as well as the confining pressure of the track. For typical track conditions, the friction angle of ballast aggregates of volcanic origin varies in the range of 45° to 65°, the upper values related to highly angular particles.
- K_0 is the coefficient of lateral earth pressure at rest in the ballast, and the lateral stresses ($\sigma_h = K_0\sigma_v$) induce lateral strains. For granular materials, the coefficient of lateral earth pressure at rest can be approximated from the effective friction angles (φ) of the materials, using the following equation:

$$K_0 = 1 - \sin\phi \tag{2.2}$$

For conventional track conditions, the friction angle of ballast is generally in the range of 45° to 60°, and this results in a value of K_0 in the range of 0.134–0.293.
- "Compressive strength" is the compressive strength of the parent rock material which is later crushed to obtain ballast particles. The compressive strength is the vertical stress (σ_1) at failure from unconfined (uniaxial) compression tests and is expressed in MPa. Cylindrical specimens for the unconfined testing are cored from the parent rock. The magnitude of the compressive strength of parent rocks depends on the type and characteristics of discontinuities or weak planes within the rock masses. For specimens with moderate discontinuities, the compressive strengths are in the range of 100–300 MPa for granite and basalt and 30–300 MPa for quartzite and limestone (Indraratna *et al.* 2011b).
- The strength is the vertical stress at failure obtained from unconfined tests and includes the density, Poisson's ratio, resilient modulus, coefficient of lateral earth pressure at rest (K_0), and effective friction angle of the ballast. Moreover, the compressive strength of the parent rock from which the ballast is made and in-track lateral confining stress to the ballast layer are also required. It should be noted that typical in-track lateral confining stresses of ballast in a conventional track environment are in the range of 10–20 kPa (Indraratna *et al.* 2017b).

2.6 Sub-ballast, subgrade/formation soils

The properties of the sub-ballast (capping layer) and subgrade formation soils are needed to enter into the program prior to the design process. Part of these parameters characterises the load-bearing performance of the sub-ballast, subgrade and includes the compacted density, Poisson's ratio, resilient modulus, coefficient of lateral earth pressure at rest and friction angle. In low-lying coastal areas in Australia, saturated clay or soft soil subgrade can become slurry or be liquefied and pumped upwards to foul the ballast under repeated train passage (Biabani *et al.* 2016b; Indraratna *et al.* 2016; Ngo *et al.* 2017b). Ballast fouling associated with clay pumping usually occurs during and after heavy rainfall and may cause the track to become unstable (Indraratna *et al.* 2011a; Ngo *et al.* 2015; Remennikov and Kaewunruen 2008). Therefore, attention must be given to drainage and filtering functions when designing sub-ballast layer (Isar *et al.* 2016; Trani and Indraratna 2010). Using geosynthetics in ballasted rail tracks may prevent or minimise ballast fouling, so this feature will be discussed and examined further in the following sections (Bathurst and Raymond 1987; Ngo and Indraratna 2016).

2.6.1 Sub-ballast/filtration layer

Sub-ballast or capping generally consists of well-graded crushed rock or sand/gravel mixtures with a usual thickness of 150 mm, placed between ballast and subgrade layers. The layer of sub-ballast acts as a filter and separating layer, transmitting and distributing imposed loads from the ballast down to the subgrade, and also acts as a drainage medium to dissipate cyclic pore water pressure (Biabani *et al.* 2016a; Nguyen *et al.* 2013). "Capping modulus" is the resilient modulus of sub-ballast/capping measured from the cyclic standard or cubical triaxial tests. The modulus is taken as the slope of the deviator stress-vertical strain (q-ε_1) curve during the unloading and reloading cycles.

2.6.2 Subgrade

The subgrade or formation layer is the platform on which the rail track structure is constructed. The subgrade may be classified into two parts: (i) natural ground (formation) and (ii) placed soil. The subgrade must be stiff and have a bearing capacity capable of supporting traffic-induced stresses at the sub-ballast/subgrade interface. The typical requirements adopted from Esveld (2001) for the subgrade and sub-ballast are presented in Table 2.2.

When a track is to be built on soft soil, the subgrade must be stabilised by ground improvement methods such as prefabricated vertical drains (PVD), lime-cement columns, deep cement/grouting and the vibratory (pneumatic) compaction or bio-engineering method of native vegetation (Indraratna *et al.* 2006, 2011b). To provide a stable platform for rail track, the subgrade must be able to prevent the failure modes given below (Selig and Waters 1994):

- Excessive progressive deformation from induced cyclic train loading
- Consolidation settlement and excessive shear failure under the combined weight of trains, track structure, earth and induced cyclic loading
- Significant changes in volume (swelling or shrinking) associated with changes of water content
- Attrition of the subgrade

Table 2.2 Typical requirements for the sub-ballast and subgrade layer

Parameters	Required values	
	Sub-ballast	Subgrade
California bearing ratio (CBR) [%]	>25	>5
E_{v2} [MPa]*	100	35
Compaction through proctor [%]	100	97
Maximum deviation from design subgrade profile [mm]	<10	<10

*E_{v2} is the modulus of elasticity taken from the second load step in a plate loading step.

There are four main input parameters for subgrade: the type of subgrade; the subgrade modulus; the thickness; and the compressive strength. It is noted from the current analysis that Li and Selig's (1998a, b) method is only applicable to the following four types of soft subgrade soils: fat clay (CH), lean clay (CL), elastic silt (MH) and silt (ML). In Li and Selig's approach, the resilient modulus of subgrade soil is limited to the range of 14–110 MPa. It should also be noted that the maximum thickness of the subgrade can be 4.5–6 m, but after 6 m the change in the deformation influence factor is less sensitive. Moreover, an increase in the resilient modulus should increase the subgrade's compressive strength.

2.7 Geosynthetics

Geosynthetics have been widely used in track construction and rehabilitation worldwide over decades (Kwon and Penman 2009). The application of geosynthetics for improving ballasted rail tracks has proved to be a cost effective way of reducing the lateral movement of ballast particles and of further reducing any permanent deformation (Brown *et al.* 2007; Ngo *et al.* 2017; Tutumluer *et al.* 2012). For ballasted railway tracks in particular, geogrids are generally used for reinforcement, which is provided by the tensile strains that develop in the geogrids and the interlocking effect between the geogrids and surrounding particles of ballast (Indraratna *et al.* 2006).

Brown *et al.* (2007) conducted a series of full-scale experiments using the composite element test to study the influence of geogrid parameters considered to be of importance to the interaction between the geogrid and ballast. The effect of the tensile strength of geogrid on the settlement of ballast was also examined. Test data indicated that the 15–65 geogrid (tensile strength =15 kN m and aperture = 65 mm) provided the least improvement, but the improvement offered by 45–65 geogrid (tensile strength = 45 kN m and aperture = 65 mm) towards the end of the test was comparable to steel mesh. They also mentioned that the cross-section of the geogrid ribs was rectangular, whereas steel mesh was circular, and it was reported that the shape of a rib was possibly a parameter affecting interaction between the geogrid and the ballast. Recently, Indraratna *et al.* (2012a) conducted a series of large-scale direct shear tests for an average size particle of ballast (d_{50}) of 35 mm, reinforced by seven geogrids with apertures varying from 20.8 mm to 80 mm, to investigate how the size of the aperture affects the shear behaviour at the ballast–geosynthetic interfaces. The laboratory results indicated that the shear strength at the interface was governed by the size of the geogrid aperture. Typical physical and mechanical properties of tested geogrids are described in Table 2.3.

Table 2.3 Physical and technical properties of geogrid and geotextile

Physical characteristics	Data
Structure	Bi-oriented geogrid
Mesh type	Rectangular apertures
Standard colour	Black
Polymer type	Polypropylene
Carbon black content	2%

(Continued)

Table 2.3 (Continued)

Physical characteristics	Data	
Dimensional characteristics	Unit	**Biaxial geogrid**
Aperture size MD	mm	40
Aperture size TD	mm	40
Mass per unit area	g m^2	420
Percentage of open area	%	77%

Technical characteristics	Unit	**Biaxial geogrid**	
		MD	TD
Tensile strength at 2% strain	kN m	10.5	10.5
Tensile strength at 5% strain	kN m	21	21
Peak tensile strength	kN m	30	30
Yield point elongation	%	11	10

Geotextile physical characteristics	Unit	Data
Mass per unit area	g m^2	140
Thickness	mm	2
Polymer type	–	Polypropylene
Geometry type	–	Non-woven

(courtesy of Polyfabrics Australia Pty Ltd)

2.8 Design criteria

General design criteria for determining the thickness of the granular layer (ballast + sub-ballast layer), such as the minimum height of ballast, the maximum allowable subgrade strain and the method of analysis, can be specified in the "design criteria" input of the program. In this section, a minimum height for the granular layer must be determined, and this commonly restricts the ballast layer to 300 mm thick and the capping layer to 150–400 mm thick, but varies the thickness of the structural fill layer, depending on subgrade conditions. As a result, the total granular thickness can be in the order of 450–1450 mm. The plastic strain allowable in the subgrade (ε_{pa}) is the total cumulative plastic strain at the surface of the subgrade for the design period expressed as a percentage, while the deformation allowable in the subgrade (ρ_a) is the total cumulative plastic deformation of the subgrade expressed in millimetres (mm). These values are only required if the "Li and Selig" method is used to calculate the capacity of the subgrade.

The "impact factor method" can also be used to calculate the impact factor (*IF*) and can also be used later to determine the design dynamic wheel load (P_d). Three widely used methods to determine *IF* are: (i) AREA (1974) method; (ii) Eisenmann (1972) method; and (iii) ORE (1965) method. The calculated design dynamic wheel load (P_d) is then used to compute the design rail seat load (q_r) using a method specified by either: (i) AREA (1974) method; (ii) ORE (1969) method; or (iii) Raymond (1977) method. These design methods can be used to determine the thickness of the granular layer. Typical input design criteria are listed below:

- Allowable subgrade plastic strain for the design period (i.e. ε_{pa} = 2%)
- Allowable settlement of subgrade in design period (i.e. ρ_a = 25 mm)
- Minimum granular layer height (i.e. H_{min} = 0.45 m)
- Impact factor method, i.e. AREA (1974) method
- Subgrade capacity method, i.e. Li and Selig (1998a, b)

2.9 Traffic conditions

"Traffic conditions" allows practising engineers to input traffic conditions that are relevant to the program for subsequent calculations. Details of these parameters are as follows:

- "Annual traffic tonnage" is a summation of all the train axle loads on the track in any one-year period. All these axle loads are measured at "one" particular location in the track. The annual traffic tonnage is expressed in "million gross tonnes" (*MGT*). The metric "tonne" is equal to 1000 kg.
- "Wheel diameter" is the average wheel diameter (expressed in metres) of all trains on the track over the design period (corresponding to the provided annual traffic tonnage). For example, the average wheel diameter of freight trains typically used in the state of New South Wales, Australia is about 0.88/0.9/0.965 m.
- "Train velocity" is the average velocity of trains travelling on the track over the design period. The track under consideration is assumed to be straight and the velocity is linear. The train velocity is expressed in kilometres per hour. Velocities of up to 160 km h^{-1} can be considered for heavy haul.
- "Static wheel load" is the average wheel load (expressed in kilo Newton – kN) of all the wagons present on the track over the design period. The static wheel load is determined when the wagon and wheel carriage are under static equilibrium in a vertical direction. Simply stated, the static wheel load can be taken as the weight of a stationary train wagon on a horizontal track divided by the number of wheels. For example, the static wheel load of a 100 t wagon having two tandem bogies is $9.81 \times 100/(2 \times 2 \times 2) = 122.6$ kN (i.e. 25 t axle load).
- "Design period" is the duration of time (expressed in years) for which the track is designed before track maintenance is required. This means the design period is used to determine the required height of the granular layer to avoid (based on the design criterion) failure or excessive settlement of the subgrade soil within that time period. The design period may be treated as variable and be used in conjunction with suitable criteria.

2.10 Rail and sleeper properties

The geometrical and mechanical properties of rails and sleepers (or ties) are determined as design input parameters. The details of these properties are explained in the following:

- Rail centre to centre spacing is the distance between the centrelines of the rails expressed in metres (G_h in Fig. 2.6). For example, rail track with a gauge of 1.435 m with 60 kg rail (which has approximately 70 mm-wide head at the point where gauge is measured) will have a centre to centre spacing of approximately 1.505 m (=1.435 + 0.070)
- The rail super elevation deficiency is the difference between the required and actual super elevations. Railways often use "cant" rather than super elevation. The difference in rail heights is called "cant" (*C* in Fig. 2.6). Cant deficiency will essentially cause an acceleration, which results when the amount of cant does not balance the speed of the train for the given curve radius (resulting in a horizontal acceleration).
- "Sleeper length" is the length of sleepers expressed in metres. The typical length of a sleeper for a standard gauge track in Australia is about 2.4–2.6 m.
- "Sleeper spacing" is the distance between two adjacent sleepers in the track expressed in metres. This distance is measured from the centre of one sleeper to the centre of another. Sleepers are generally spaced at distances (centre-to-centre) in the range of 0.6–0.75 m.

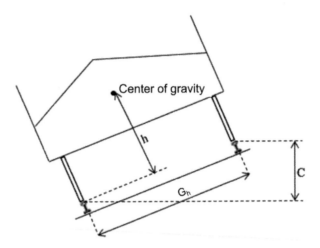

Figure 2.6 Schematic representation of parameters *C, h* and *G$_h$* of a train car travelling along a curved section of track

- "Sleeper width" is the width of a sleeper expressed in metres (i.e. at the base of sleeper). Widths of 0.20–20.25 m are common for sleepers currently used in Australia.
- The vertical distance between the top of the rail and the centre of mass of the vehicle is the perpendicular distance from the top of the two rails to the centre of mass (m) of the vehicle, and is denoted by *h* in Figure 2.6.

Chapter 3

Bearing capacity of ballasted tracks

3.1 Introduction

The bearing capacity of a ballast module allows one to determine the bearing capacity of ballast based on the limit equilibrium method (Le Pen and Powrie 2010). Prior to designing the track substructure, the bearing capacity of the selected ballast must be checked to ensure it meets the required design criteria. This module enables track engineers to decide whether or not to carry out a full design process, depending on the bearing capacity required.

The mechanical parameters of the granular layer play a governing role in determining its bearing capacity. The unit weight of ballast can vary from 14 to 18 kN m^{-3}. The friction angle of ballast is the effective friction angle at the *in situ* confining pressure; this plays a governing role in determining the allowable bearing capacity of the ballast layers. It has to be noted that as ballast is highly non-linear, the friction angle can vary with the confining pressure that prevails in the field. For conventional track conditions, the friction angle of ballast is between 45° and 55°, but the maximum value can be up to 65° depending on size and angularity of particles. If the designers need to determine the dynamic bearing capacity, Vesic (1973) recommends obtaining the input friction angle by subtracting 2° from the static friction angle of the ballast.

3.2 Calculation of design wheel load (P)

There are several empirical formulae that can be applied to calculate the vertical wheel loads (*P*) required. They are all expressed as a function of the static wheel load (P_s) and the impact factor (*IF*):

$$P = IF.P_s \tag{3.1}$$

Note that the impact factor *IF* is dimensionless and its value is always greater than the unity regardless of the formulae used (AREMA 2003). Also note that the static wheel load is determined by the gross weight of a train car divided by the total number of wheels of the car.

3.2.1 AREA (1974) method

The American Railroad Engineering Association AREA (1974) introduced a simple mathematical expression to calculate the value of *IF* based on the results of *in situ* measurements of dynamic wheel loads from train cars with known static wheel loads. All the measurements were undertaken on standard gauge tracks (1435 mm) in the US. The value of *IF* is set to be a function of vehicle speed *V* (km h) and wheel diameter *D* (mm):

$$IF = 1 + 5.21\frac{V}{D} \tag{3.2}$$

3.2.2 Eisenmann (1972) method

Eisenmann's method is based on a statistical analysis of the actual measurement of dynamic wheel loads. Complete description of the formulation of this model is also given in Jeffs and Tew (1991). The value of the impact factor *IF* is calculated as follows:

$$IF = 1 + \beta \times \delta \times \eta \times t \tag{3.3}$$

The factor β takes into account the different dynamic performance of loaded and unloaded vehicles as follows:

$\beta = 1.0$ for loaded vehicles

$\beta = 1.5$ for empty vehicles

$\beta = 1.3$ for locomotives with unsprung masses of about 3.5 tonnes per axle

The effect of track condition is included using the factor δ, the value of which can be taken from the five track conditions listed as follows:

$\delta = 0.1$ for track in very good condition

$\delta = 0.2$ for track in good condition

$\delta = 0.3$ for track in average condition

$\delta = 0.4$ for track in poor condition

$\delta = 0.5$ for track in very poor condition

η is the speed factor and is computed using the following expressions:

$$\eta = \frac{V}{60} \qquad \text{for vehicle speeds up to 60 km/h}$$

$$\eta = 1 + \frac{V - 60}{140} \qquad \text{for vehicle speeds in the range of 60 to 200 km/h}$$

The parameter *t* is determined from the upper confidence limit that is the probability that the maximum dynamic wheel load is not exceeded. Eisenmann (1972) analysed an extensive collection of *in situ* dynamic wheel load data and recommended the following values:

$t = 0$ for upper confidence limit of 50%

$t = 1$ for upper confidence limit of 84.1%

$t = 2$ for upper confidence limit of 97.7%

$t = 3$ for upper confidence limit of 99.9%

3.2.3 ORE (1969) method

The Office of Research and Experiments (ORE) of the International Union of Railways has proposed a more comprehensive formulation to determine the value of the dimensionless impact factor. The formulation of IF is based entirely on the results of the measured track of locomotives (ORE 1969) and is expressed by:

$$IF = 1 + \alpha' + \beta' + \gamma' \qquad (3.4)$$

The parameter α' depends on vertical track irregularities and vehicle suspension and speed, but for practical design purposes, correlating the value of α' with track irregularities can be difficult. Therefore, ORE introduced a simple equation based on actual track experiments to approximate the value of α'. In the least accurate case, increases of α' can be modelled with a cubic function of the vehicle velocity, V (km h):

$$\alpha' = 0.04 \left(\frac{V}{100} \right)^3 \qquad (3.5)$$

The effects of curved tracks on the vertical components of dynamic wheel loads are accounted for by the parameter β', which is expressed by:

$$\beta' = \frac{2dh}{G_h^2} \qquad (3.6)$$

The parameter d is the super elevation deficiency (m). The perpendicular distance from the two rails to the vehicle's centre of mass (m) is denoted by h. G_h is the horizontal distance between the centreline of the rails (m), g is the gravitational acceleration (m sec^{-2}), and V is the speed of the vehicle (km h^{-1}).

The track and vehicle conditions are accounted for in a formulation through the parameter γ' where in a manner similar to α' the value of γ' is, in the least accurate case, approximated as a cubic function of the velocity of the vehicle (km h^{-1}), thus:

$$\gamma' = 0.1 + 0.017 \left(\frac{V}{100} \right)^3 \qquad (3.7)$$

3.3 Calculation of maximum rail seat load

The value of the design wheel load (P) obtained from the previous section is used to calculate the maximum rail seat load (q_r). There are three models to determine the value of q_r as described below.

3.3.1 AREA (1974) method

AREA (1974) developed a relationship to determine q_r based on the beam of the elastic foundation model, but this particular reference is implicit because the value of q_r is calculated from:

$$q_r = D_f \times P \qquad (3.8)$$

Where D_f is the distribution factor whose value depends on the spacing of the sleeper S_s (expressed in mm) and the type of sleeper:

$$D_f = 0.40 + 5.77 \times 10^{-4} s_s \quad \text{for timber sleepers}$$
$$D_f = 0.33 + 6.73 \times 10^{-4} s_s \quad \text{for steel sleepers}$$
$$D_f = 0.45 + 5.77 \times 10^{-4} s_s \quad \text{for concrete sleepers}$$

3.3.2 ORE (1969) method

Base on a statistical analysis of experimental results, ORE (1969) introduced an empirical formula to estimate the maximum rail seat load as:

$$q_r = \bar{\varepsilon} \times c_1 \times P \tag{3.9}$$

The value of c_1 can be taken as approximately 1.35, while $\bar{\varepsilon}$ can best be approximated from the following table.

3.3.3 Raymond (1977) method

This method follows recommendations by Raymond (1977) and is based on an assumption that the maximum possible loading occurs when the wheel load on a rail is distributed between three adjacent sleepers. The maximum load is assumed to be $x\%$ of the wheel load, and this takes place at the rail seat of the central sleeper when the wheel load P is directly above it. The rail seat loads on the other two sleepers are assumed to be $x/2$. This simple assumption results in a maximum wheel load of:

$$q_r = 0.5P \tag{3.10}$$

Table 3.1 Values of the ratio $\bar{\varepsilon}$ for various types and spacing of sleepers; the area of sleeper support = 2Q×sleeper width at rail seat, and Q = distance from the centreline of the rail to the end of the sleeper

Sleeper type	Sleeper dimensions (mm) length×width×rail seat thickness	Sleeper spacing (mm)	Area of sleeper support (10^3 mm²)	Q (mm)	$\bar{\varepsilon}$
French, type VW pre-stressed Concrete	2300 × 250 ×140	600	200	400	0.58
British, type F pre-stressed Concrete	2515 × 264 × 200	760	269	510	0.40
German, type 58 pre-stressed Concrete	2400 × 280 × 190	600	252	450	0.38
French, hardwood	2600 × 255 × 135	600	281	550	0.61

3.4 Calculation of ballast/sleeper contact pressure

The maximum rail seat load is then used to determine the contact pressure at the sleeper and ballast interface (P_a). The general approach for this calculation is to assume a uniform distribution of contact pressure over the estimated "effective area" of the sleeper (A). Neglecting the weight of the sleeper, the contact pressure is the maximum rail seat load per unit effective area. A factor of safety (F_2) is usually included in the calculation to account for any uncommon distribution of contact pressure and other irregularities. AREA method has been adopted in this book.

3.4.1 AREA (1974) method

To calculate the contact pressure P_a, the AREA method assumes that the maximum rail seat load is twice as large as the value obtained from the equations in the previous section, and the effective area is assumed to be the total area of the sleeper in contact with the ballast (Jeffs and Tew 1991).

$$P_a = \left(\frac{2q_r}{Bl}\right)F_2 \tag{3.11}$$

Typical values of F_2 between 2 and 3 have been recommended by various rail organisations. Maximum sleeper-ballast contact pressures have also been introduced to prevent damage to different types of sleepers. As one of the most widely accepted method, AREA (1974) has recommended a maximum contact pressure of 450 kPa and 590 kPa for timber and concrete sleepers, respectively.

3.5 Bearing capacity of ballast

In this section. the limit equilibrium approach for determining the bearing capacity of rail track is presented. The ultimate bearing capacity of ballast q_{ult} can be obtained using the following equations (Le Pen and Powrie 2010):

$$q_{ult} = N_\gamma S_\gamma \left(0.5\,\gamma B - \Delta u\right) \tag{3.12}$$

$$N_q = K_p e^{\pi \tan \phi} \tag{3.13}$$

$$K_p = \left(\frac{1+\sin\phi}{1-\sin\phi}\right) \tag{3.14}$$

$$N_\gamma = \left(N_q - 1\right)\tan\left(1.4\phi\right) \tag{3.15}$$

$$S_\gamma = 1 + 0.1K_p\left(B/L\right) \tag{3.16}$$

where γ is the bulk unit weight of ballast, Δu is the pore water pressure increment, L is the length of the sleeper, B is the width of the sleeper, φ is the angle of effective shearing resistance of ballast and N_q, N_γ and S_γ are standard bearing capacity factors.

In order to account for the dynamic condition, Vesic (1973) proposed the use of the dynamic friction angle (φ_{dy}) in Equation 3.17, for the above equations 3.12–3.16.

$$\phi_{dy} = \phi - 2° \tag{3.17}$$

The allowable bearing capacity is thus calculated by:

$$Allowable\ bearing\ capacity = \frac{Ultimate\ bearing\ capacity\ (q_{ult})}{FOS} \tag{3.18}$$

Chapter 4

Thickness of granular layer

4.1 Introduction

In conventional design methods, the ballast height needed to spread the load to an acceptable level is determined based on a trial height and iterative calculations, but these methods mostly ignore the subgrade properties, ballast degradation, deformation and other important factors. This chapter introduces the "thickness of granular layer" module, which allows the designer to calculate the required thickness (or height) of the granular layers overlying relatively weaker subgrade soils which are prone to bearing capacity-type shear failure and excessive settlement. Specific types of subgrade soils which belong to this category include silty and clayey soils. The granular layers in a typical track comprise the ballast, sub-ballast and structural fill layers (Fig. 4.1).

4.2 Procedure to determine the thickness of ballast and capping layer

This section presents the steps and procedure to determine the thickness of ballast and capping layers based on two design criteria: (i) preventing excessive subgrade deformation and failures, and (ii) limiting the vertical stress on the subgrade soils to less than "threshold stress" in order to protect against subgrade failure by excessive plastic deformation. The procedure used to design the granular layer thickness is illustrated in Figure 4.2. The threshold stress is calculated from repeated load tests in which the cumulative strain of the subgrade is measured as a function of the number of loading cycles applied. Once the applied stress level exceeds the threshold stress, the rate of cumulative plastic deformation of the subgrade will be very fast, whereas at applied stress levels below the threshold stress level, the rate of the cumulative plastic deformation is slow. The three inputs of the resilient modulus of ballast, capping and structural fill must be determined prior to calculating the thickness of the granular layer. These input moduli are taken into calculation separately, based on the assumption that it is constant over the full height of the granular layer in each calculation (i.e. Li and Selig method).

4.2.1 Design procedure

Step 1: Impact factor (*IF*) calculation: AREA (1974) method is used:

$$IF = 1 + 5.21\frac{V}{D} \qquad (4.1)$$

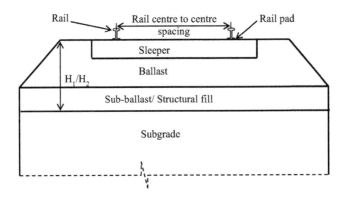

Figure 4.1 Track composite structure consisting of sleeper, ballast, sub-ballast and subgrade

Step 2: Dynamic wheel load (P_d) calculation:

$$P_d = IF \times P_s \tag{4.2}$$

Step 3: Number of load cycle (N) calculation:

$$N = \frac{T}{8 \times P_s} \tag{4.3}$$

where T is the total traffic tonnage and P_s is the static wheel load.

Step 4: Select the values of a, b and m for designed soil type from Table 4.1 (e.g. for CH soil; $a = 1.2$; $m = 2.4$; $b = 0.18$)

Step 5: Calculation for the first design procedure (preventing local shear failure of subgrade)

Step 5.1: Calculate allowable deviator stress on subgrade, σ_{da} (kPa):

$$\sigma_{da} = \sigma_s \left[\left(\frac{\varepsilon_{pa}}{aN^b} \right)^{(1/m)} \right] \tag{4.4}$$

where σ_s is soil compressive strength (kPa) and ε_{pa} is allowable subgrade plastic strain for the design period (i.e. $\varepsilon_{pa} = 2\%$)

Step 5.2: Calculate the strain influence factor (I_ε)

$$I_\varepsilon = \frac{\sigma_{da} A}{P_d} \tag{4.5}$$

where A is an area factor arbitrarily selected to make the strain influence factor dimensionless. Its value is taken as 0.645 m². P_d is the design dynamic wheel load. The value of P_d

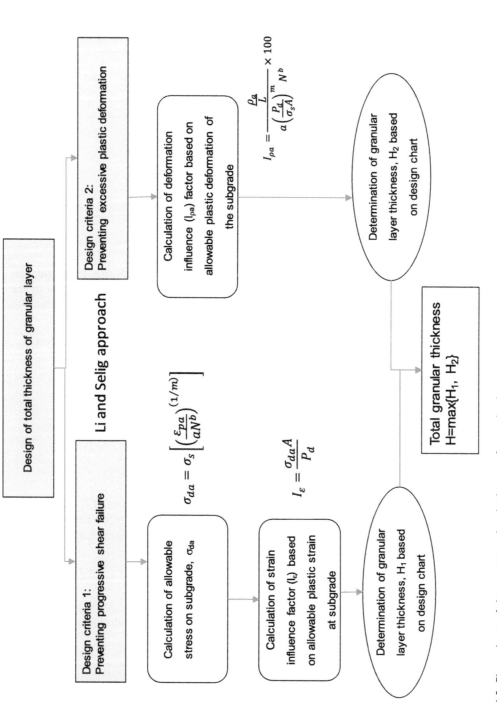

Figure 4.2 Flow chart of determining the thickness of granular layer

Table 4.1 Values of soil parameters a, b and m for calculating the plastic shear strain of subgrade for four types of soil

Soil type	a	b	m
CH (fat clay)	1.20	0.18	2.40
CL (lean clay)	1.10	0.16	2.00
MH (elastic silt)	0.84	0.13	2.00
ML (silt)	0.64	0.10	1.70

(adopted from Li and Selig 1998a)

is calculated from the static design wheel load P_s using the American Railway Engineering Association (AREA) recommended equation (AREA 1974):

$$P_d = \left(1 + \frac{5.21V}{D}\right) P_s \qquad (4.6)$$

Step 5.3: Determine the (H/L) from Figure 4.3 that corresponds to a given granular material modulus and designed subgrade modulus, and then the thickness of the granular layer can be obtained, H_1 (note that L is the length factor, L = 0.152 m). For a complete set of design charts, users may refer to Li et al. (1996).

Step 6: Calculation for the second design procedure (preventing excessive plastic deformation of the subgrade layer)

Step 6.1: Determine the deformation influence factor I_p:

$$I_\rho = \frac{\dfrac{\rho_a}{L}}{a\left(\dfrac{P_d}{\sigma_s A}\right)^m N^b} \times 100 \qquad (4.7)$$

Step 6.2: Determine the (H/L) from Figure 4.4 based on the values of I_p, and the type of soil, and then the thickness of the granular layer (H_2) can be obtained (note that L is the length factor, L = 0.152 m). For a complete set of design charts, users may refer to Li et al. (1996).

The design thickness of the granular layer (H) is determined based on the larger value (obtained from Step 5.3 and Step 6.2), that is $H = \max[H_1, H_2]$.

It is noted that the above procedure is realistic when the modulus of the granular layer is a single entity, i.e. the modulus of ballast. This is in accordance with the analysis of the granular thickness described by Li and Selig (1998a, b).

4.3 Equivalent modulus and strain analysis

This section provides track engineers with an option to predict an equivalent modulus of granular layer once the following input parameters have been determined (Fig. 4.5): elastic modulus of ballast; elastic modulus of capping; elastic modulus of structural fill.

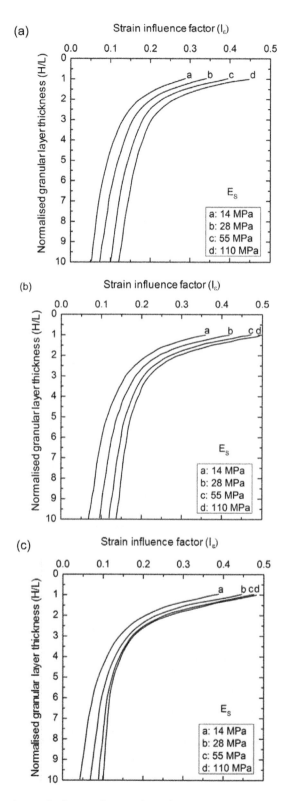

Figure 4.3 Granular layer thickness design chart for preventing progressive shear failure:
(a) E_b = 550 MPa, (b) E_b = 280 MPa and (c) E_b = 140 MPa

(modified after Li and Selig 1998a – with permission from ASCE)

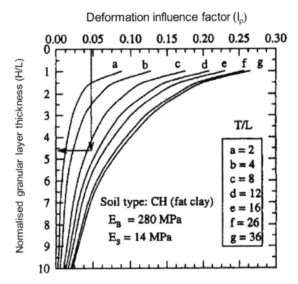

Figure 4.4 Granular layer thickness design chart for preventing excessive plastic deformation (modified after Li and Selig 1998b – with permission from ASCE)

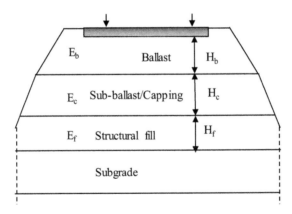

Figure 4.5 Schematic diagram of a typical ballasted track embankment

The modulus of granular layer ($\bar{E}$) of the whole track substructure (Fig. 4.5) can be estimated by:

$$\bar{E} = \frac{H_b + H_c + H_f}{\left(\dfrac{H_b}{E_b} + \dfrac{H_c}{E_c} + \dfrac{H_f}{E_f} \right)} \tag{4.8}$$

where, E_b, E_c and E_f are the elastic modulus of the ballast, capping and structural fill, respectively. H_b, H_c and H_f are the thicknesses of ballast, capping and structural fill, respectively.

The average strain (ε_{ave}) of the equivalent granular medium is calculated as the stress at sleeper/ballast interface divided by the equivalent modulus ($\bar{E}$). The elastic settlement ($S_{elastic}$) of granular medium is then predicted using the total height:

$$\varepsilon_{ave} = \frac{q_r / \left(B \times \dfrac{L}{3} \right)}{\bar{E}} \tag{4.9}$$

$$S_{elastic} = H \times \varepsilon_{ave} \tag{4.10}$$

where q_r is rail seat load (kN) determined by the AREA (1974) method presented earlier.

Vertical stress on the subgrade can be calculated using three different methods, i.e. the trapezoidal approximation, Talbot and Boussinesq methods.

Vertical stress on subgrade can be determined as (i.e. trapezoidal method):

$$q_d = \left[\frac{q_r}{(B+H) \times (L/3+H)} \right] \tag{4.11}$$

where q_r is the rail seat load and H is selected total thickness of granular layer.

The allowable bearing capacity is then calculated as the ultimate bearing capacity divided by the factor of safety (FOS). The ultimate bearing capacity is expressed as:

$$q_{ult} = (\pi + 2) q_{uc} \tag{4.12}$$

$$q_{allowable} = q_{ult} / FOS = (\pi + 2) \times q_{uc} / FOS \tag{4.13}$$

where q_{uc} is the undrained shear strength of the subgrade.

The selected granular layer thickness is appropriate if $q_{allowable} > \sigma_v$; vertical stress on subgrade.

4.4 Determination of track modulus

4.4.1 Introduction

The current world-wide trend towards increased axle loads and faster heavy haul trains has resulted in exacerbated damage to tracks. The ability to accurately assess the structural condition of track through appropriate health monitoring schemes has become crucial. One important parameter for characterising the track substructure condition is to evaluate accurately the track stiffness or modulus using *in situ* or large-scale prototype testing.

The traditional way of assessing the track modulus is to measure the first deflection without a load when stationary, and then increase the load to obtain a load-deflection curve at a given location on the track (Selig and Waters 1994). It is noted that designers often do not know to a reliable extent the magnitude of track deflection without actually measuring it. Therefore, it is a dilemma for practitioners as the use of an incorrect track modulus can result in an unrealistic track deflection and substructure deformation (Indraratna *et al.* 2011b).

Previous studies have suggested that measurement of track deflection under an applied vertical load may be used to assess the track structural conditions, namely the average track stiffness that includes the entire layering of the substructure. In contrast, the resilient modulus of each track element with depth is considered to be an important parameter, and the overall modulus is seldom measured and its magnitudes are at best approximately known for most sections of the track (Selig and Waters 1994). Therefore, it is imperative to have an accurate method for determining track modulus to improve railroad design and prediction of degradation under cyclic train loading.

A number of theoretical models have been proposed for the calculation of track modulus based on load–deflection relationships, yet there is no consensus on the best or the most accurate method. The most commonly known method assumes the track assembly as a beam on an elastic foundation (i.e. the Winkler model), as illustrated in Figure 4.6. A vertical force (P) applied by an axle produces a vertical rail deflection (w). Therefore, the track stiffness (k), taken at a point as the wheel (bogie) passes directly above it, is defined as follows:

$$k = \frac{P}{w} \tag{4.14}$$

where k = track stiffness; P = vertical force applied by a wheel; w = measured deflection of the rail.

Ebersöhn and Selig (1994) showed that a load-deflection test for a selected load increment can be used to determine *in situ* track stiffness as follows:

$$k = \frac{P}{w} = \frac{P_f - P_0}{w_f - w_0} \tag{4.15}$$

where P_f, P_0 are the final and initial vertical rail force, respectively; and w_f, w_0 are the final and initial rail elevation, respectively.

From this, the track modulus (u) can be determined as a function of the stiffness from:

$$u = \frac{k^{4/3}}{(64EI)^{4/3}} \tag{4.16}$$

where E is the Young's modulus of the rail and I is the moment of inertia of the rail. It is noted that the difference between track modulus (u) and track stiffness (k) is that k includes the rail bending stiffness EI, whereas u is only related to the remainder of the superstructure (i.e., the fasteners and sleepers) and the substructure (ballast, sub-ballast, and subgrade).

While this method is theoretically sound, it has a number of practical shortcomings.

- The method does not take into account the initial closing of voids in the track upon application of the load (i.e. known as the "seating" modulus). In addition, it does not consider the representation of individual substructural layers (i.e. ballast, capping and subgrade formation).
- By considering only a selected point on the track, the values obtained are subject to variation due to local inconsistencies at the test site. Taking a measurement at a single point may not be representative of the entire track.

It is clear that the track modulus and stiffness are directly interrelated. The only difference is that the modulus is independent of the rail properties and should not change with the change of rail stiffness, whereas the stiffness includes the effect of the rail, including geometry and material modulus. However, the modulus as well as the stiffness will change with a

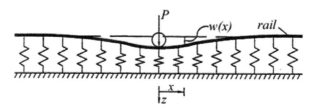

Figure 4.6 Typical rail-in-track subjected to a wheel load

change in the type of sleeper and its spacing as placed on track. Therefore, designers should be mindful of when to use this approach, thus:

- If the purpose of measuring the track load-deflection characteristics is to obtain the track modulus to be used with the Winkler method for track design, then all the superstructure properties and the changes in properties along the length of the track should be recorded.
- If the purpose is to examine the support condition (substructure layers of tracks), the variation in support is more important than obtaining the track modulus. Using the track stiffness or measuring the variations in track deflection under a set of constant loads gives a direct indication of the change in support conditions.

Alternative approaches to determining track modulus may also be considered, as suggested by past studies from North America and Europe assuming that the rail-supporting base (consisting of pads, ties, ballast and subgrade) may be represented by a system of springs for each layer with a different stiffness, arranged in series (Fig. 4.7). In this approach, the rail support modulus for the entire base is given by:

$$k = \frac{1}{\dfrac{1}{k_p} + \dfrac{1}{k_t} + \dfrac{1}{k_b} + \dfrac{1}{k_c} + \dfrac{1}{k_s}} \tag{4.17}$$

where k_p is the corresponding stiffness of the pad (if used), k_t is the stiffness of the tie (due to the compressibility of timber/concrete in the rail-seat region and tie bending), k_b is the vertical stiffness of the ballast layer and k_s is the stiffness of the subgrade.

Using this approach, the overall track stiffness can be obtained as long as the stiffness of ballast, capping and subgrade are available (Equation 4.17). Although intuitively appealing field tests to determine the stiffness/modulus values of track components are necessary, if those values are obtained using laboratory tests, then the corroborating responses for stiffness values based on ballast and subgrade test specimens must be quantified for accurate track deformation analysis. Naturally, the error in track deformation resulting through this approach is mainly attributed to the assumption of the ballast layer being elastic, when it is actually elasto-plastic and even yielding under high loads and imparting particle degradation (Indraratna *et al.* 2011b; Lackenby *et al.* 2007).

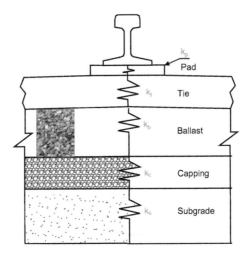

Figure 4.7 Schematic of typical track structure

4.5 Determining the resilient modulus of ballast, M_R

A number of large-scale triaxial tests have been conducted on fresh and clay-fouled ballast to determine the resilient modulus of ballast, M_R. Details of these laboratory tests and testing procedures can be found elsewhere, as described by Tennakoon (2012), Lackenby *et al.* (2007) and Sun *et al.* (2014a). The resilient modulus of ballast, M_R can be determined as:

$$M_R = \frac{\Delta_{q,cyc}}{\varepsilon_{a,rec}} \qquad (4.18)$$

$$\Delta_{q,cyc} = q_{max} - q_{min} \qquad (4.19)$$

where $\Delta_{q,cyc}$ is the magnitude of deviator stress and $\varepsilon_{a,rec}$ is the recoverable portion of axial strain.

Based on extensive laboratory tests carried out by Tennakoon (2012), the resilient modulus M_R of fresh and fouled ballast is presented in Figure 4.8. It can be observed that M_R increases with an increase of the loading cycles, N at a diminishing rate, until it approaches a constant value. An increase in the level of fouling results in a decrease in M_R. For instance, this would lead to an enforcement of speed restrictions.

A relatively low value of M_R also indicates the occurrence of higher elastic strains compared to clean ballast, as the deviator stresses are generally maintained as constant during train loading. This implies that fouling increases both the elastic and plastic strain components. More elastic strains result in higher track vibration, which may further decrease the resiliency of ballast (low M_R).

4.5.1 Empirical relationship to determine resilient modulus

An attempt was made by Lackenby (2006) and Indraratna *et al.* (2009) to establish an empirical relationship to determine the resilient modulus of ballast, as described below.

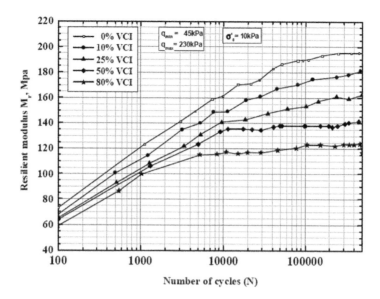

Figure 4.8 Resilient modulus, M_R, of fresh and fouled ballast

(data source: Tennakoon 2012)

Indraratna *et al.* (2009) introduced a hyperbolic model, sometimes cited as the K-ϕ model (Hicks, 1970). This model suggests M_R as a function of the bulk stress φ (i.e. sum of principal stresses, $\phi = \sigma_1' + \sigma_2' + \sigma_3'$):

$$M_R = k_1 \phi^{k_2}$$ (4.20)

where k_1 and k_2 are empirical coefficients.

A series of large-scale triaxial tests have been conducted by Indraratna *et al.* (2009), and they found that all values of M_R fall within a narrow band, irrespective of σ_3 and q_{max}. For latite ballast, this highlights the emergence of a unique relationship between the resilient modulus and bulk stress, given by:

$$M_R = 40 \phi^{0.34}$$ (4.21)

Sun (2015) proposed an empirical equation to predict M_R of ballast with frequency (*f*) and bulk stress (φ) as follows:

$$M_R = m \times f^n + \phi^t$$ (4.22)

where *m*, *n* and *t* are empirical parameters, and typical values of resilient modulus for ballast are given in Figure 4.9.

Lackenby (2006) proposed an empirical equation relating the four parameters resilient modulus M_R, number of cycles N, maximum deviator stress $q_{max,cyc}$ and effective confining pressure σ_3'. This is an extension of the model suggested by Brown (1974), where the number of load cycles, N, is included.

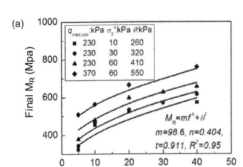 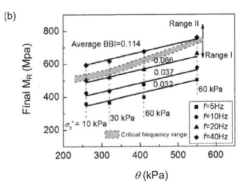

Figure 4.9 Resilient modulus M_R response under cyclic loading
(data source: Sun 2015)

Table 4.2 G and H values for $q_{max,\,cyc}$ = 500 kPa

N	G	H
100	267.9	−0.201
1000	343.4	−0.176
2000	373.9	−0.155
3000	369.0	−0.124
5000	387.8	−0.114
10000	426.4	−0.129
25000	439.4	−0.108
50000	465.4	−0.108
100000	467.4	−0.097
250000	478.0	−0.096
500000	466.5	−0.070

(data source from Lackenby 2007)

$$M_R = G \left(\frac{q_{max,cyc}}{\sigma_3'} \right)^H \qquad\qquad (4.23)$$

where G and H are parameters that will depend on the number of load cycles and $q_{max,cyc}$ is as listed in Table 4.2; the measured and calculated values of M_R are presented in Table 4.3.

4.5.2 Measured field values of dynamic track modulus

Nimbalkar and Indraratna (2016) studied the track modulus at discrete points along the track during normal scheduled train operations, through an extensive field trial in the town of Singleton, New South Wales (NSW), Australia. Four types of geosynthetics and a rubber shock mat were installed below the ballast layer in selected sections of track constructed on three different subgrades (soft alluvial clay, hard rock and concrete bridge), and the performance

Table 4.3 Measured and calculated coefficients G and H for $q_{max, cyc}$ = 500 kPa at 2000 loading cycles

σ_3' (kPa)	N	M_R (MPa) (measured)	M_R (MPa) (predicted)	$\Sigma error^2$
10	2000	210	204	36
20	2000	226	227	1
30	2000	253	242	129
45	2000	246	258	137
60	2000	253	269	251
90	2000	294	287	47
120	2000	300	300	0
180	2000	318	319	1
240	2000	339	334	29

(data sourced from Lackenby 2006)

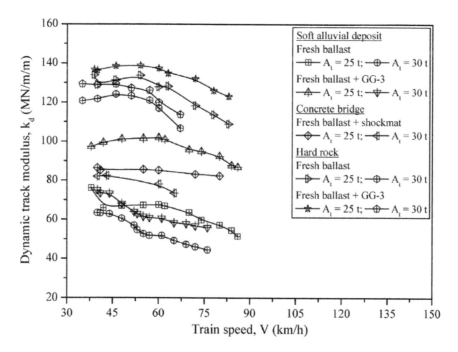

Figure 4.10 Dynamic track modulus as a function of train speed

(data source: Nimbalkar and Indraratna 2016 – with permission from ASCE)

of the instrumented track was monitored for five years under in-service conditions, including tamping operations. The method used in this study is in accordance with the theoretical approach described by Priest and Powrie (2009), whereby the measurement of loading magnitude and frequency relevant to in-service conditions could be readily obtained. Figure 4.10 shows that the calculated track modulus varies significantly from one section to another and also illustrates the role of subgrade type and the favourable effects of geogrids. It is noted that, unlike the conventional approach, the dynamic amplification of a static wheel load has been considered in this analysis. The track modulus for the fast train is about 40–60% less than for the slow trains, which is much larger than that obtained by Esveld (2001) and Yang *et al.* (2009) on the basis of the dynamic magnification effect alone.

Values of track modulus presented in Figure 4.10 can be used in preliminary track design. It is recommended that laboratory or field tests be carried out to obtain more reliable and accurate values of track modulus for a complete ballasted track design.

Chapter 5

Effect of confining pressure and frequency on ballast breakage

5.1 Introduction

Railway tracks are mainly influenced by the degradation of ballast particles (e.g. Ferreira and Indraratna 2017; Indraratna *et al.* 2011b; Lobo-Guerrero and Vallejo 2005; Lu and McDowell 2006; Sun *et al.* 2016, among others). Particle breakage under dynamic loads is a complex mechanism that usually begins at the inter-particle contacts (i.e. breakage of asperities), followed by a complete crushing of weaker particles under further loading. A rapid fragmentation of particles and subsequent clogging of voids with fines is commonly observed in overstressed railway foundations (Huang *et al.* 2009; Powrie *et al.* 2007). The degradation of particle is the primary cause of contamination and accounts for up to 40% the fouled material (Dombrow *et al.* 2009; Feldman and Nissen 2002).

The main factors that affect ballast degradation can be grouped into three categories: (i) properties related to the characteristics of the parent rock (e.g. hardness, specific gravity, toughness, weathering, mineralogical composition, internal bonding and grain texture); (ii) physical properties related to individual particles (e.g. soundness, durability, particle shape, size, angularity and surface smoothness); and (iii) factors associated with the assembly of particles and loading conditions (e.g. confining pressure, initial density or porosity, thickness of ballast layer, ballast gradation, the presence of water or ballast moisture content, and a cyclic loading pattern including load amplitude and frequency). While the properties of individual grains of ballast such as size, shape and angularity govern its degradation under traffic loading, deformation is also influenced by the magnitude of wheel (or axle) load, frequency (equivalent to train speed) and the number of load cycles (Sun *et al.* 2014a, 2016). This chapter presents the influences of confining pressure and frequencies on the degradation of railway ballast.

5.2 Determination of ballast breakage

Several methods have been used to quantify particle breakage; some of these indices are presented in this section.

Hardin (1985) introduced a relative breakage index of:

$$B_r = \frac{B_t}{B_p} \tag{5.1}$$

where B_t and B_p are the total breakage and breakage potential, respectively. The potential for a particle of soil to break increases with size because the normal contact forces in a soil element increase with particle size, as does the probability of micro-cracks in a given particle increasing with size. Therefore, the breakage of particles of soil in a sample of rockfill under moderate stresses will be quite evident, whereas very high stresses are needed to crush silt size particles.

Marsal (1973) introduced a breakage index B_g to quantify the breakage of rock-fill materials. His method involves changing the overall grain-size distribution of aggregates after a load has been applied and then carrying out a sieve analysis on specimens before and after each test. From the changes recorded in particle gradation, the difference in percentage retained on each sieve size ($\Delta W_k = W_{ki} - W_{kf}$) is computed whereby W_{ki} represents the percentage retained on sieve size k before the test and W_{kf} is the percentage retained on the same sieve size after the test. Marsal (1967) noticed that some of these differences are positive and some negative; in fact, the sum of all positive values of ΔW_k must be theoretically equal to the sum of all negative values. Marsal (1967) defined the breakage index B_g as the sum of the positive values of ΔW_k, expressed as a percentage, where the breakage index B_g has a lower limit of zero, which indicates no particle breakage, and a theoretical upper limit of the unity (100%) which represents a situation with all the particles broken to sizes below the smallest sieve size used.

Lade et al. (1996) proposed a new particle crushing parameter with permeability correlations in mind, based on the D_{10} particle size distribution, as given by:

$$B_{10} = 1 - \frac{D_{10f}}{D_{10i}} \tag{5.2}$$

where B_{10} = particle breakage factor; D_{10f} = effective grain size of the final gradation; and D_{10i} = effective grain size of the initial gradation. This particle breakage factor is formulated based on the lower limit being zero when there is no particle breakage and the upper limit being the unity at infinite particle breakage.

Indraratna et al. (2005) introduced an alternative ballast breakage index (BBI) based on particle size distribution (PSD) curves. They reported that previous triaxial testing on ballast indicates that particle degradation causes a shift in the initial particle size distribution towards smaller particle sizes, while the maximum size does not change much before and after loading. Therefore, instead of defining the breakage potential by a single minute particle, they considered an arbitrary boundary of maximum breakage to be more appropriate. The ballast breakage index (BBI) is determined on the basis of a change in the fraction which passes a range of sieves, as shown in Figure 5.1. This increase in the degree of breakage causes the PSD curve to shift towards the smaller particle size region on a conventional PSD plot. Thus, if the area A (between the initial and final PSD) increases, the amount of ballast breakage is high. It is noted that the BBI has a lower limit of 0 (no breakage) and an upper limit of 1. By referring to the linear particle size axis, the BBI can be calculated using the following equation:

$$BBI = \frac{A+B}{A} \tag{5.3}$$

where A is the area defined previously, and B is the potential breakage or area between the arbitrary boundary of maximum breakage and the final particle size distribution.

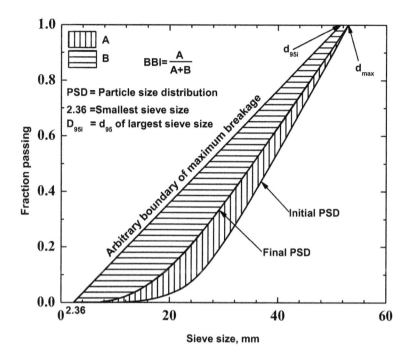

Figure 5.1 Ballast breakage index (BBI) calculation method
(modified after Indraratna *et al.* 2013b – with permission from ASCE)

5.3 Influence of confining pressure on ballast breakage

Lackenby *et al.* (2007) described two problems arising from increasing axle loads: differential track settlement and ballast degradation. One potential method of enhancing the substructure is to manipulate the level of ballast confinement. Although the influence that the confining pressure has on various geotechnical structures is an important design criteria, it is commonly underestimated in conventional rail track design and construction. To investigate this possibility, a series of high frequency cyclic triaxial tests were carried out at the University of Wollongong (i.e. Indraratna *et al.* 2011b; Sun *et al.* 2016) to examine the effects of the magnitudes of confining pressure and deviator stress on ballast deformation (permanent and resilient) and degradation using a series of large-scale testing apparatus designed and built at the University of Wollongong, Australia.

5.3.1 Prototype testing and experimental simulations

The conventional triaxial apparatus is a versatile method for obtaining the deformation and strength of coarse- and fine-grained materials in the laboratory, but the discrepancy between the actual particle shapes and sizes in the field and the reduced particle sizes adopted in conventional laboratory equipment leads to inaccurate load-deformation responses and failure modes. These inaccurate measurements are the results of inevitable size-dependent dilation and different mechanisms of particle crushing. This means that testing coarse aggregates in conventional apparatus can give misleading results because of disparities between particle and equipment size, but as the ratio of the specimen to maximum particle size exceeds 6, the effects due to equipment boundaries can be ignored (Indraratna *et al.* 1998). To mitigate these size dependent issues, large-scale facilities for testing ballast have been designed and built at the University of Wollongong over the last two decades, as shown in Figure 5.2.

A large-scale triaxial testing apparatus (Fig. 5.2a), which can accommodate samples of 300 mm in diameter and 600 mm high, was used for testing railway ballast. The apparatus consists of five main parts: the triaxial chamber, the vertical loading unit, the oil reservoir and pump, the servo-control unit and the digital data acquisition system. The deviator stress was applied by a dynamic actuator capable of frequencies up to 60 Hz at load amplitudes of 150 kN. Each specimen was subjected to a given loading frequency, and the loading was suspended after 500,000 cycles irrespective of the axial strain achieved. Before and after loading, the ballast was passed through the set of 12 standard sieves (2.36–53 mm) at least twice to ensure accurate breakage estimation.

A large-scale process simulation testing apparatus (PSTA) was designed and built to study the response of ballast track components under realistic cyclic loading (Fig. 5.2b); it was the first of its kind in the world when used in the early 1990s. This PSTA can accommodate specimens 800 mm long, 600 mm wide and 600 mm high; these dimensions were selected to mimic a typical unit cell section of Australian standard gauge tracks. Furthermore, this PSTA is a true triaxial apparatus because three independent principal stresses can be applied in three mutually orthogonal directions, and it can apply a 100 kN dynamic actuator load with frequency of up to 40 Hz to simulate heavy haul trains with a 40-tonn axle load travelling at up to 300 km/hour.

A large permeameter has also been designed (Fig. 5.2c) to measure the hydraulic conductivity of ballast contaminated with fouling materials such as coal and subgrade mud; this apparatus is 0.5 m in diameter and 1 m high. A filter membrane was placed above a coarse granular layer (prepared from coarser ballast aggregates) while still maintaining a free drainage boundary to prevent fouling material flowing out. The thickness of ballast layer in Australian rail track varies between 300 and 500 mm. In view of this, a 500 mm thick ballast layer was used to determine the permeability of fouled ballast. The test specimen was placed above the filter membrane and compacted in four equal layers to represent a typical field density of 1600 kg m³. Commercial kaolin (plastic and liquid limits are 26.4% and 52.1%, respectively) was used to simulate the clay fouling. A predetermined amount of fouling corresponding to different degrees of fouling was mixed with ballast and compacted to gain a similar density of ballast, so that the voids of the ballast were kept constant throughout the test series.

High-capacity drop-weight impact testing equipment (Fig. 5.2d) was also used to examine the effects of rubber mats in the attenuation of dynamic impact loads and subsequent mitigation of ballast degradation (Kaewunruen and Remennikov 2010). The impact apparatus consists of a free-fall hammer of 5.81 kN that can be dropped from a maximum height of 6 m, guided through rollers on the vertical columns fixed to the strong floor. These drop heights and drop mass were selected to produce dynamic stresses in the range of 400–600 kPa, thus simulating a typical wheel-flat and dipped rail joint. The impact loads are monitored by a dynamic load cell (capacity of 1,200 kN), mounted on the drop hammer and connected to a computer-controlled data acquisition system. An accelerometer is attached to the top surface of ballast assembly to measure accelerations during the impact tests.

It is also noted there are several limitations with these testing devices, including: (i) inability to simulate the moving load that actually occurs in the field; (ii) difficulties in simulating a

(a) Large-scale triaxial apparatus

(b) Process simulation testing apparatus
     (PSTA)

(c) Large-scale permeability test
     apparatus (photo taken at HighBay
     lab – UOW)

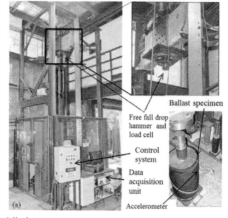

(d) Impact testing apparatus (modified
     after Kaewunruen and Remennikov
     2010)

Figure 5.2 The large-scale ballast testing equipment at the University of Wollongong,
Australia

high number of load cycles due to boundary constraints; (iii) some boundary conditions will
affect the test results, particularly large ballast particles.

5.3.2 Laboratory study on the effect of confining pressure on ballast degradation

A series of cyclic triaxial tests were carried out to investigate how confining pressures
affect ballast under cyclic loading (e.g. Indraratna *et al.* 2011b; Lackenby *et al.* 2007; Ngo
et al. 2015; Sun *et al.* 2016, among others). Ballast specimens could be prepared to the

recommended gradation (i.e. $d_{50} = 38.5$ mm, $C_u = 1.54$, $e_o = 0.76$; where d_{50} is the diameter of ballast that corresponds to 50% finer in the particle size distribution curve and C_u is the coefficient of uniformity) and then tested under effective confining pressures (σ_3') ranging from 1 to 240 kPa with $q_{max} = 500$ kPa. The results indicate that for each deviator stress considered, an "optimum" range of confining pressures exists such that degradation is minimised. This range varies from 15 to 65 kPa for a maximum deviator stress of 230 kPa and increases up to 50–140 kPa when the deviatoric stresses increase to 750 kPa. Ballast specimens tested at low confining pressures indicative of current *in situ* conditions were characterised by excessive axial deformations, volumetric dilation and an unacceptable degree of degradation associated mainly with angular corner breakage. The results suggest that *in situ* lateral pressures should be increased to counteract the axle loads of heavier trains and that practical methods of achieving increased confinement should be adopted.

Figure 5.3a shows the results of confining pressure (σ_3') on the axial and volumetric strains of ballast measured at the end of 500,000 cycles; here the axial strains decrease with an increasing confining pressure and the ballast assemblies dilate at a low confining pressure ($\sigma_3' < 30$ kPa), but they become progressively more compressive as the confining pressure increases from 30 to 240 kPa. The effect of confining pressure on particle breakage is shown in Figure 5.3b, where breakage is divided into three regions: (I) a dilatant unstable degradation zone (DUDZ); (II) an optimum degradation zone (ODZ); and (III) a compressive stable degradation zone (CSDZ). The data show that the specimens are subjected to rapid and considerable axial and expansive radial strains that result in an overall volumetric increase or dilation at a low confining pressure in the DUDZ region ($\sigma_3' < 30$ kPa). Due to the low confining pressures in this zone, the ballast particles have limited particle-to-particle areas of contact, but as the confining pressure increases to the ODZ region ($\sigma_3' = 30$–75 kPa), the rate of axial strain decreases due to an apparent increase in stiffness, and the overall volumetric behaviour is slightly compressive (Lackenby *et al.* 2007). Therefore, the particles in this region are held together in an optimum array with enough lateral confinement to provide an optimum distribution of contact stress and increased areas of inter-particle contact, all of which reduce the risk of particle breakage.

As σ_3' increases further in the CSDZ region ($\sigma_3' > 75$ kPa), the particles rub against each other, which limits their sliding and rolling but significantly increases their breakage. Increasing confinement decreases track settlement while increasing track stability and stiffness (resilient modulus) under cyclic loading, leading to greater track stability and passenger comfort. In summary, this study indicates the optimum confining pressure based on loading and track conditions and suggests that track confinement can be increased by decreasing the spacing of sleepers, increasing the height of shoulder ballast, including a geosynthetic (i.e. geogrids, geocomposite) at the ballast–sub-ballast interface, widening the sleepers at both ends and using intermittent lateral restraints at various parts of the track (Indraratna *et al.* 2011b).

5.3.3 Prediction of axial strains and ballast breakage

Based on extensive cyclic laboratory tests carried out at the University of Wollongong using large-scale equipment, the estimation of permanent axial strain $\varepsilon_{a,p}$ and ballast breakage (*BBI*) after 500,000 load cycles, under varying confining pressures, $\sigma_3 = 10$–240 kPa subject to cyclic loading with $q_{max} = 230$ kPa, 500 kPa and 750 kPa presented in Figures 5.4 and 5.5. It is noteworthy that subject to repeated train loading, the ballast aggregate breaks and causes

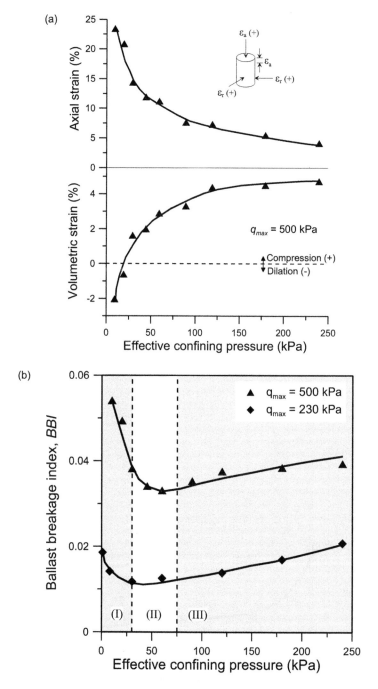

Figure 5.3 Effect of confining pressures on: (a) axial and volumetric strains; (b) particle degradation (after Lackenby *et al.* 2007)

the ballast to settle (i.e. increased permanent axial strains). Figure 5.4 shows that permanent axial strains can be expressed as a function of effective confining pressure and ballast breakage index BBI (based on regression analysis of the laboratory test data). This strain is then used after the total ballast height has been obtained to calculate the height of ballast needed to ensure that after the specific amount of breakage, its height will still be sufficient. It is given in the following formula:

$$H_{required} = \frac{H_{initial}}{100 - \varepsilon_{a.p}} * 100 \qquad (5.4)$$

The settlement of ballast can therefore be estimated as:

$$S = H_{required} - H_{initial} \qquad (5.5)$$

This change is applicable to all methods of calculating ballast height.

5.3.4 Resilient modulus of ballast

In the Li and Selig method of determining ballast height presented in Chapter 4, the resilient modulus of the ballast is an important factor in determining the final height. Each design chart used by Li and Selig is each formulated for a specific value of ballast resilient modulus. Based on UOW test data, it is found that the resilient modulus (M_R) of ballast can be empirically represented as a function of ballast confining pressure (σ_3) and breakage (BBI). In addition, laboratory test data showed that the resilient modulus M_R of the ballast layer increases as

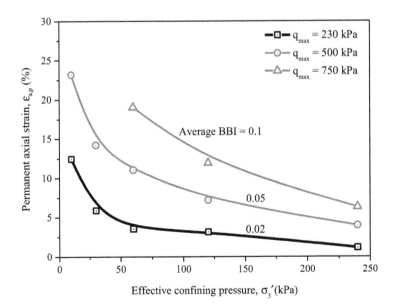

Figure 5.4 Effect of confining pressure on ballast breakage

(data source from Lackenby *et al.* 2007)

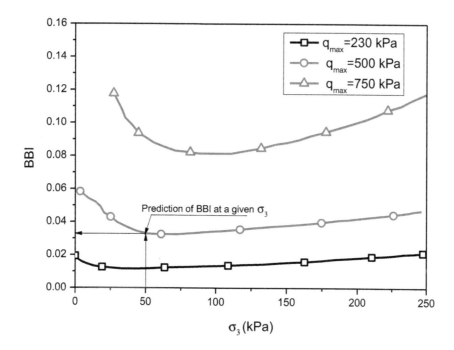

Figure 5.5 Prediction of ballast breakage, BBI, at a given σ_3 based on UOW test data (modified after Lackenby *et al.* 2007)

the confining pressure increases, and the ballast breakage index, BBI (i.e. at maximum cyclic stress), is also found to critically affect the resilient modulus, as presented in Figure 5.6. This graph provides empirical relationships, allowing practicing engineers an option considering the influence of confining pressures on the resilient modulus in track design.

Ballast breakage can also be accounted for in calculating the thicknesses of ballast and sub-ballast layers by means of the ballast breakage index (*BBI*), as explained by Indraratna *et al.* (2011b). The value of *BBI* can be obtained independently from the UOW test data and manually entered into the module of input parameter (Chapter 2). This calculated resilient modulus will be used later in a subsequent calculation of the thicknesses of the layers of ballast and sub-ballast. If users consider the influence of ballast breakage on settlement for a given track design, then an additional ballast vertical strain will be included as a result of particle breakage (Fig. 5.4). The value of this additional strain is used to determine a sacrificial thickness of ballast, which is later used to determine the recommended heights of the ballast and sub-ballast layers, as given in Equation 5.4. Alternatively, the user can manually enter preferred values for the vertical strain from breakage and effective resilient modulus (i.e. refer to Figs 5.3–5.6).

5.4 Influence of frequency on ballast breakage

The influence of train speed on the permanent deformation and degradation of ballast during cyclic loading has been studied using the large-scale cylindrical triaxial apparatus (Fig. 5.7a).

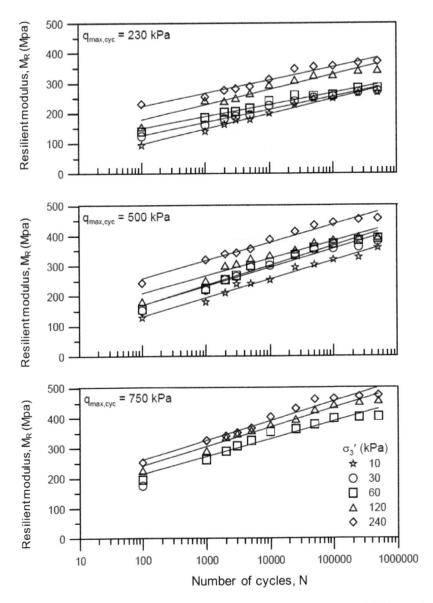

Figure 5.6 Measured resilient modulus of ballast at a given σ_3 based on UOW test data (data source from PhD thesis, Lackenby 2006)

These test specimens were isotropically consolidated to confining pressures (σ_3') of 10, 30 and 60 kPa, and then frequencies (f) varying from 5 Hz to 60 Hz were used to simulate train speeds from about 40 to 400 km h; maximum cyclic deviator stresses ($q_{max,cyc}$) of 230 and 370 kPa were then used to represent axle loads of 25 and 40 tonnes, respectively.

Figures 5.7a and 5.7b presents the variation of axial strain (ε_a) versus the number of cycles (N) for different frequencies (f) and load amplitudes ($q_{max,cyc}$) of cyclic loading. For a specific

f, the ε_a increases with the increase of N, but for different f, ε_a increases as f increases at a specific N. There are four regimes of permanent deformation based on the cyclic loading applied: (i) a zone of elastic shakedown with no accumulation of plastic strain, (ii) a zone of plastic shakedown characterised by a steady state response and a small accumulation of plastic strain, (iii) a ratcheting zone with a constant accumulation of plastic strain, and (iv) a plastic collapse zone where plastic strains accumulate rapidly and failure occurs in a relatively short time (Sun *et al.* 2016). Moreover, three different deformation ranges

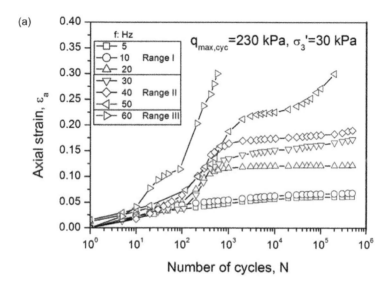

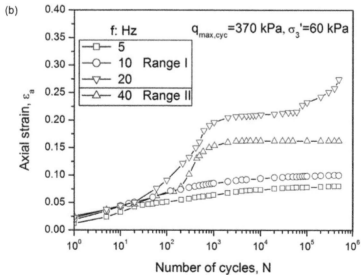

Figure 5.7 (a) Variation of axial strain ε_a verse number of cycles N for $q_{max,cyc}$ of 230 kPa; (b) variation of axial strain ε_a versus number of cycles N for $q_{max,cyc}$ of 370 kPa; (c) variation of ballast breakage index BBI with various frequencies f; (d) examples of ballast degradation

(Sun *et al.* 2016 – with permission from ASCE)

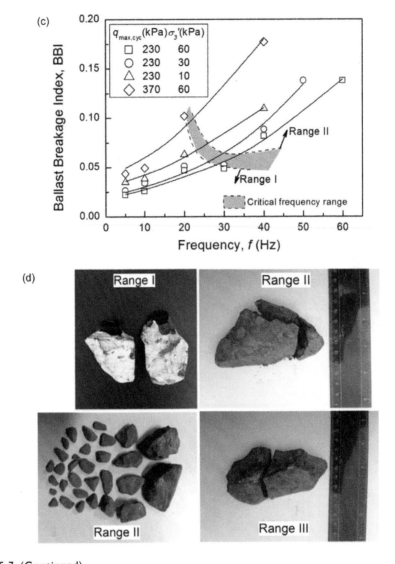

Figure 5.7 (Continued)

appeared in response to the loading frequency, namely: Range I–Plastic shakedown at $f \leq 20$ Hz (or $V \leq 145$ km h), Range II–Plastic shakedown and ratcheting at 30 Hz $\leq f \leq 50$ Hz (or 220 km h $\leq V \leq 360$ km h), and Range III–Plastic collapse at $f \geq 60$ Hz (or $V \geq 400$ km h), as shown in Figures 5.7a and 5.7b. Cyclic triaxial data on ballast from Lackenby *et al.* (2007) indicates that similar regimes of permanent deformation response exist based on the applied stress ratio ($q_{max,cyc}p$').

A range of critical frequency is identified as 20–30 Hz for σ_3' = 30 kPa and 30–40 Hz for σ_3' = 60 kPa, respectively. Figure 5.7c shows that critical frequency decreases as particle breakage increases, and ratcheting failure (Range II) of the specimen would occur with

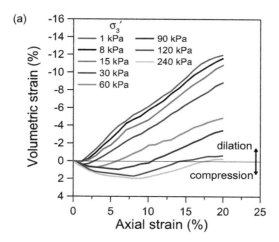

(a)   Monotonic loading

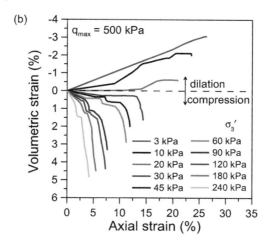

(b)   Cyclic loading

Figure 5.8 Volumetric strains of ballast tested under monotonic and cyclic loading
(data source: Indraratna *et al.* 1998; Lackenby 2006 – with permission from ASCE)

significant particle breakage (BBI > 0.10), even at a relatively low value of frequency (i.e. $f = 25$ Hz). Distinct ballast degradation which corresponds to different ranges of deformation occurred during cyclic testing; in Range I ($f \leq 30$ Hz), particle degradation took the form of attrition of asperities and corner breakage (Fig. 5.7d), but as the frequency increased (30 Hz $< f < 60$ Hz) in Range II, particle splitting caused by fatigue and a high degree of attrition resulting from increased vibration became predominant. At a very high frequency ($f \geq 60$ Hz) in Range III, the coordination number is greatly reduced, which would induce particle splitting.

5.5 Volumetric behaviour of ballast under monotonic and cyclic loading

Following extensive laboratory tests, the volumetric strains measured at different confining pressures are shown in Figure 5.8, where dilation (volume increase) occurred in the ballast samples for most confining pressures under monotonic loads. Although similar ballast materials were also tested in the same large-scale triaxial apparatus, they all exhibited different volumetric strain responses under different loading conditions. However, those ballast assemblies which underwent cyclic loads (i.e. confining pressure higher than 30 kPa) experienced pronounced compression, possibly due to the reorientation and rearrangement of particles which occurs during cyclic loading and generates a denser (compressing) or looser (dilating) packing assembly. Those specimens subjected to low confinement exhibited purely dilative behaviour, whereas the reverse occurred for assemblies with higher confining pressures.

Chapter 6

Impact of ballast fouling on rail tracks

6.1 Introduction

During operation, ballast deteriorates due to the breakage of angular corners and sharp edges, infiltration of fines from the surface and mud pumping from the subgrade under train loading, as illustrated in Figure 6.1. As a result of these actions, ballast becomes fouled and less angular, and its shear strength is reduced (Indraratna *et al.* 2011a; Ngo *et al.* 2016). Fouling materials have traditionally been considered as unfavourable to track structure. The sources, quantifications and adverse effects of fouling are discussed in the following sections.

Fouling is caused by a number of mechanisms associated with traffic loads and various maintenance cycles, such as tamping, ballast-cleaning and reconstruction processes. Selig and Waters (1994) stated that ballast can become fouled from various sources and categorised them into five main groups, as given in Table 6.1. These sources of fouling can be divided into three main scenarios: the first case is due to the degradation of ballast, infiltration of fines dropped from wagons and particles transported by wind or water. The second fouling mechanism is fine particles in the layers of sub-ballast (sand, crushable cinders or slag) that are degraded and injected into the ballast under traffic loads. The third cause of ballast fouling is subgrade mud pumping, which is attributed to water mixing with particles of subgrade (clay or soft soils) to form a clay slurry, which pumps up through voids in the overlying ballast layer. To prevent this type of fouling, a layer of sub-ballast with proper gradation or geosynthetics placed on top of the subgrade was considered to be an effective method to prevent the formation of slurry by eliminating any attrition of the subgrade and inhibiting the slurry from migrating into the ballast voids (Ngo *et al.* 2018; Rujikiatkamjorn *et al.* 2012).

6.2 Quantifying of ballast fouling

There are several indices for quantifying the degree of ballast fouling. Selig and Waters (1994) defined the fouling index (*FI*) as a summation of percentage (by weight) passing through a 4.75 mm and 0.075 mm sieve. This method may lead to misinterpretation of the actual degree of fouling if the fouled material contains more than one type of material with different specific gravities (e.g. coal and pulverised rock). Alternatively, Feldman and Nissen (2002) defined the percentage void contamination (*PVC*) as the ratio of bulk volume of fouling material (V_{vf}) to the initial voids volume of clean ballast (V_{vb}):

$$PVC = \frac{V_{vf}}{V_{vb}} \times 100 \qquad (6.1)$$

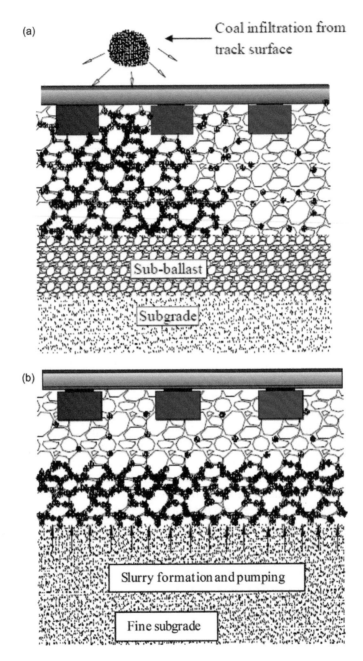

Figure 6.1 Typical fouling mechanism in ballasted tracks
(after Tennakoon *et al.* 2012)

Table 6.1 Sources of ballast fouling

I Ballast breaks down

   a Handling of ballast

      i At quarry
      ii During transportation
      iii From dumping

   b Thermal stress from heating (desert)
   c Freezing of water in particles
   d Chemical weathering (including acid rain)
   e Tamping damage
   f Traffic damage

      i Repeated load
      ii Vibration
      iii Hydraulic action of slurry

   g From compaction machines

II Infiltration from ballast surface

   a Delivered with ballast
   b Dropped from trains
   c Wind blown
   d Water borne
   e Splashing from adjacent wet spots
   f Meteoric dirt

III Sleeper (tie) wear
IV Infiltration from underlying granular layers

   a Old track bed breakdown
   b Sub-ballast particle migration from inadequate gradation

V Subgrade infiltration

Although the *PVC* method is a direct measure of percentage of voids occupied by fouling particles, the measurement of volume is time consuming and the parameter V_{vf} must be calculated after compacting the fouling material (standard Proctor technique), which does not always represent the actual volume of fouling accurately in a track environment. Most fouling indices are based on mass ratios, none of which represent the influence or the degree of void clogging. Moreover, when ballast is contaminated by different types of fouling material, these methods may not give an accurate assessment because they do not consider of the specific gravity of fouling materials that can be much smaller than the ballast. According to Feldman and Nissen (2002), the bulk volume of fouling material (V_f') must be determined based on the specimen compacted at a standard Proctor energy level. The compacted bulk volume does not always represent the actual volume of fouling in a track environment. In view of the above, the void contaminant index (*VCI*) proposed by Tennakoon *et al.* 2012 and Indraratna *et al.* (2013b) is described below:

$$VCI = \frac{V_{f'}}{V_{vb}} \times 100 \qquad (6.2)$$

where V_f' is the actual volume of fouling material within the ballast voids. By substituting the relevant soil parameters, Equation 6.2 can be re-written as:

$$VCI = \frac{1+e_f}{e_b} \times \frac{G_{s.b}}{G_{s.f}} \times \frac{M_f}{M_b} \times 100 \tag{6.3}$$

where e_f = void ratio of fouling material, e_b = void ratio of fresh ballast, G_{sb} = specific gravity of ballast, G_{sf} = specific gravity of fouling material, M_f = dry mass of fouling material and M_b = dry mass of fresh ballast.

The advantage of Equation 6.3 is that it includes different types of fouling materials such as coal, mud or pulverised ballast and incorporates their respective values of specific gravity. For this reason, the void contaminant index (*VCI*) is used in this research to quantify ballast fouling.

For example, a *VCI* = 50% indicates that half of the total voids in the ballast are occupied by fouling material. The effect of fouling on geotechnical characteristics such as permeability and shear strength depend on the type of fouling materials (e.g. coal vs. clay), so a proper understanding of the nature of fouling materials is pertinent irrespective of the quantity of fouling. For example, sand and coal fouling may not decrease the overall permeability of the track significantly, whereas clay fouling can reduce track drainage more dramatically (Indraratna *et al.* 2011b; Rujikiatkamjorn *et al.* 2013; Selig and Waters 1994).

6.3 Relation among fouling quantification indices

The laboratory tests used to measure their FI, PVC and VCI values were carried on clay-fouled ballast (simulated by kaolin as the fouling material), sand-fouled ballast (simulated with clayey fine sand as fouling material) and coal-fouled ballast. Figure 6.2 shows the comparison between FI, PVC and VCI for various percentages of fouling. For instance, let us consider 15% fouling by mass for coal-fouled, clay-fouled and sand-fouled ballast, where the corresponding VCI values are 78%, 65% and 52%, respectively. The associated values of FI for these different percentages are 16, 28 and 15, respectively. It is clear that coal-fouled and sand-fouled ballast give a very close value to each other (i.e. difference of 16–15 = 1) in spite of the difference in the specific gravities of coal and sand (quartz), compared to the difference in VCI (78–52 = 26). The PVC values for the three fouling materials are 54%, 48% and 42%, but these three values are not as widespread (42–54%) compared to the range of VCI values (52–78%). Therefore, VCI is more sensitive to changes in the type and extent of fouling, apart from being more realistic, because it is the only method of characterising fouling that incorporates the specific gravity of the fouling material.

The initial placement density of the ballast in the actual rail track is often ascertained as standard practice in most countries. Most Australian standards for ballast (AS 2758.7 1996; TS 3402, 2001) recommend the range of *in situ* densities of ballast. While the authors agree that these can vary in the field depending on the tamping efforts, they can still be considered as reasonable estimates. Ballast degradation also contributes quite substantially to ballast fouling, which in turn justifies the need for a more rational parameter, such as VCI, which can consider the effect of types of fouling material such as coal, clay, sand and mineral filler that result from ballast breakage. The need for additional laboratory tests such as specific gravity, moisture content and a proper field sampling procedure should be encouraged in order to avoid costly track maintenance works, which are often governed by an inaccurate assessment of fouling based on mass-based fouling indices such as FI.

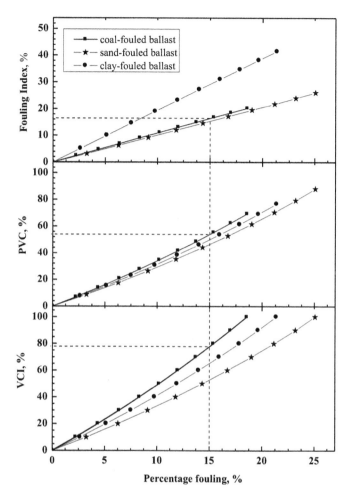

Figure 6.2 Comparison between fouling index, percentage void contamination and void contaminant index for various ranges of percentage of fouling

(data source: Tennakoon 2012)

6.4 Influence of ballast fouling on track drainage

Track substructure should be designed and constructed so as to drain the water into nearby drainage ditches or pipes. Internal drainage is usually ensured by placing a layer of sub-ballast with an appropriate gradation. The primary purpose of drainage is to remove water from the substructure of the track as quickly as possible and keep the load bearing stratum as dry as possible. To accomplish this, the load bearing layer (ballast) is usually composed of coarse and uniformly graded aggregates with large voids that ensure a sufficiently high permeability. Since the ballast is laid on fine grained subgrade, a filtering layer (sub-ballast) is usually placed below the ballast in order to prevent the upward ingress of fines.

6.4.1 Drainage requirements

To design a satisfactory drainage system, it is imperative to examine the conditions of the subsurface, groundwater and climate. Subsurface investigations must be carried out to characterise the subgrade soils, including type, layering and permeability. The proposed drainage system should have sufficient capacity to drain the highest expected rate of water during the designed life of the system. The first requirement is to keep the ballast clean enough to ensure a sufficiently high permeability for rapid drainage (Selig and Walter 1994). Secondly, the surface of the sub-ballast and subgrade should be sloped towards the sides. The third requirement is to provide a suitable means (channel or conduit) of carrying the water away which emanates from the substructure.

As mentioned previously, an increasing degree of fouling significantly reduces the drainage capacity.

For fairly uniform sand, Hazen (1911) proposed an empirical relationship as:

$$k = cD_{10}^2 \tag{6.4}$$

where
k = coefficient of permeability (cm s^{-1})
c = empirical constant
D_{10} = effective size (mm)

A theoretical formulation for the coefficient of permeability generally referred to the Kozeny-Carman equation is given by:

$$k = \frac{1}{C_s S_s^2 T^2} \frac{\gamma_w}{\mu} \frac{e_3}{(1+e)} \tag{6.5}$$

where
k = coefficient of permeability
C_s = shape factor
S_s = surface area per unit volume
T = Tortuosity
γ_w = unit weight of water
μ = absolute coefficient of viscosity
E = void ratio

Those models work well for some types of granular materials such as sands and silts, but with coarse-grained aggregate such as ballast, with its larger and inter-connected pore structure, the change of hydraulic conductivity with respect to the porosity is usually insensitive unless a large amount of fines such as coal and clay has accumulated in the voids. In order to represent the hydraulic conductivity (k) of a mixture of granular soil and fine grained soil, Koltermann and Gorelick (1995) proposed:

$$k = \frac{d_{fp}^2 \phi_{fp}^3}{180(1-\phi_{fp})^2} \tag{6.6}$$

where ϕ_{fp} is the composite porosity of the mixture and d_{fp} is the representative diameter of the grain.

The above model fails to represent fouled ballast in the track because it assumes the fine particles to be distributed uniformly throughout the voids, whereas in the field, fouling materials increasingly accumulate towards the bottom of the ballast layer (i.e. vertical migration under vibration and rainfall ingress and subsequent compaction upon the passage of trains).

Tennakoon *et al.* (2012) studied the effect of the degree of fouling on the overall hydraulic conductivity of fouled ballast using a large-scale permeability apparatus. In this research, Darcy's law was adopted as the hydraulic gradients were low enough to maintain the linear regime. An equivalent hydraulic conductivity for ballast mixed with the contaminating fines (e.g. coal fines or clay) can be obtained as:

$$k = \frac{k_b \times k_f}{k_f + \frac{VCI}{100} \times (k_b - k_f)} \tag{6.7}$$

where k_b and k_f are the values of hydraulic conductivity of clean ballast and fouling material, respectively. VCI is the void contaminant index described in section 6.2 and is expressed as a percentage.

A relative hydraulic conductivity ratio (k_b/k) was introduced to understand how the overall hydraulic conductivity (k) varies in comparison to clean ballast (k_b). This indicates how many times the overall hydraulic conductivity of fouled ballast may be reduced compared to clean ballast. Equation 6.7 can then be rearranged as:

$$\frac{k_b}{k} = 1 + \frac{VCI}{100}\left(\frac{k_b}{k_f} - 1\right) \tag{6.8}$$

In order to maintain cost-effective ballast cleaning operations, the drainage conditions in the field should be categorised with respect to the degree of fouling. Based on the current observations, and inspired by previous literature (Selig and Waters 1994; Terzaghi and Peck 1967), the drainage effect of fouled ballast with respect to the hydraulic conductivities is presented in Table 6.2. These values are then tabled in terms of relative hydraulic conductivities, as shown in Table 6.3. An example of variations of relative hydraulic conductivity with VCI and corresponding drainage criterion adopted from Terzaghi and Peck (1967) is plotted in Figure 6.2. It should be noted that drainage descriptors for railway ballast are subjective, and they will vary depending on the local climate, track use, etc. For example, a hydraulic conductivity of 10^{-5} m s^{-1} for a section of track may be acceptable due to the low rainfall etc., but unacceptable for a track section subjected to very heavy rainfall.

Table 6.2 Drainage criteria of fouled ballast

Hydraulic conductivity/ (m s^{-1})	10^0	10^{-1}	10^{-2}	10^{-3}	10^{-4}	10^{-5}	10^{-6}	10^{-7}	10^{-8}
Drainage condition	Free drainage		Very good drainage		Good drainage	Acceptable drainage	Poor drainage		Very poor drainage	Impervious

Table 6.3 Drainage condition criteria based on relative hydraulic conductivity

Drainage condition	Relative hydraulic conductivity
Free drainage	$k_b/k \leq 3$
Very good drainage	$3 < k_b/k \leq 30$
Good drainage	$30 < k_b/k \leq 300$
Acceptable drainage	$300 < k_b/k \leq 3000$
Poor drainage	$3000 < k_b/k \leq 300,000$
Very poor drainage	$300,000 < k_b/k \leq 30,000,000$
Relatively Impervious	$k_b/k \leq 30,000,000$

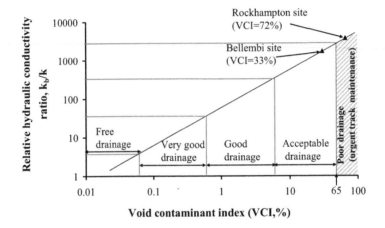

Figure 6.3 Variation of relative hydraulic conductivity with void contaminant index (%) for coal-fouled ballast

(data source from Tennakoon PhD thesis, 2012)

6.4.2 Fouling versus drainage capacity of track

A series of constant head permeability tests were carried out at the University of Wollongong using a large-scale permeameter (Fig. 6.4) to measure the hydraulic conductivity associated with different levels of fouling and to establish a relationship between the void contaminant index (VCI) and associated permeability. This apparatus is 0.5 m in diameter and 1 m high. A filter membrane was placed above a coarse granular layer (prepared from coarser ballast aggregates) while still maintaining a free drainage boundary to prevent fouling material flowing out. The thickness of ballast layer in most rail tracks varies between 300 and 500 mm, so 500 mm thick ballast layer was used to determine the permeability of fouled ballast. The test specimen was placed above the filter membrane and compacted in four equal layers to represent a typical field density of 15.5 kN m^{-3}. Commercial kaolin (plastic and liquid limits are 26.4% and 52.1%, respectively) was used to simulate the clay fouling. A predetermined amount of fouling corresponding to different degrees of fouling was mixed with ballast and compacted to gain similar density of ballast, so that the voids of the ballast (V_1) were kept constant throughout the test series.

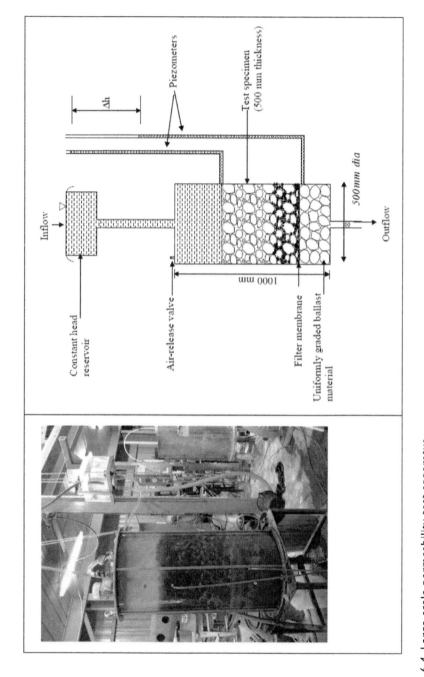

Figure 6.4 Large-scale permeability test apparatus
(photo taken at HighBay Lab-UOW)

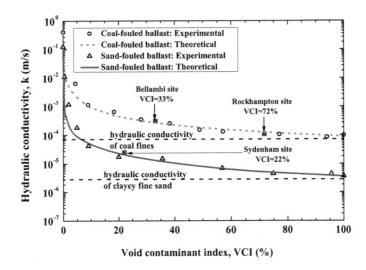

Figure 6.5 Variation for hydraulic conductivity vs. void contaminant index for coal/clayed fine sand

As expected, the overall hydraulic conductivity always decreases with an increase in VCI (Fig. 6.5). The test results show that a 5% increase of VCI decreases the hydraulic conductivity by a factor of at least 200 and 1500 for ballast contaminated by coal and fine clayey sand, respectively. However, this reduction in permeability would not affect the minimum drainage capacity needed for acceptable track operation. Beyond VCI of 75%, further reduction in hydraulic conductivity becomes marginal as it approaches the hydraulic conductivity of the fouling material itself. These observations agree with the laboratory measurements of sand–gravel mixtures reported by Jones (1954), whereby a high percentage of sand (greater than 35%) in gravel would provide a hydraulic conductivity close to that of the sand itself.

Figure 6.6 shows the variation of hydraulic conductivity for clay (pure kaolin) fouled ballast where the fouling material is uniformly distributed. At small levels of VCI, the overall hydraulic conductivity of ballast is relatively unaffected, but beyond VCI = 90%, the overall permeability of fouled ballast is almost the same as kaolin.

6.5 Fouling versus operational train speed

Based on UOW laboratory data on coal and clay-fouled ballast, the following empirical relationship representing the normalised shear strength of fouled ballast can be proposed (Indraratna *et al.* 2013b):

$$\frac{q_{peak,f}}{q_{peak,b}} = \frac{1}{1+\beta\sqrt{(VCI)}} \tag{6.9}$$

where $q_{peak,b}$ and $q_{peak,f}$ are peak deviator stresses for fresh and fouled ballast, respectively, and β is an empirical parameter for fouled ballast. The values of β are given in Table 6.4 for clay- and coal-fouled ballast tested at UOW. For other fouling types, it is suggested that users need to perform laboratory tests (i.e. large-scale triaxial test) to obtain beta parameters.

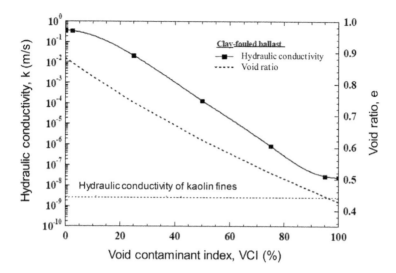

Figure 6.6 Variation hydraulic conductivity with void contaminant index for uniform clay fouled ballast

Table 6.4 Values of β based on test data

Confining pressure, σ'_3, kPa	Peak deviator stress $q_{peak,b}$, kPa	β	
		For clay-fouled ballast	For coal-fouled ballast
10	280	0.094	0.031
30	340	0.047	0.025
60	470	0.050	0.042

Adopting the American Railway Engineering Association (AREA 1974) for the design wheel load P_d and Raymond's (1977) method for calculating the rail seat load q_r, the train speed V_f for fouled ballast is obtained as:

$$V_f = \left[\frac{q_{peak,f} \times (B+h/2)(l/3+h/2)}{0.5 \times P_s} - 1 \right] \frac{D_w}{0.0052} \tag{6.10}$$

B = sleeper width (m); l = sleeper length (m); h = ballast thickness (i.e. 0.3m); D_w is the diameter of the wheel (m), and V is the velocity of the train (km h)

For fresh ballast (i.e. $VCI = 0$, $q_{peak,f} = q_{peak,b}$), train speed (V_b) is expressed as:

$$V_b = \left[\frac{q_{peak,b} \times (B+h/2)(l/3+h/2)}{0.5 \times P_s} - 1 \right] \frac{D_w}{0.0052} \tag{6.11}$$

A velocity reduction ratio (VRR) is then introduced as:

$$VVR = \frac{V_f}{V_b} \tag{6.12}$$

Substituting for V_f and V_b from Equations 6.10 and 6.11, respectively, VRR is obtained as:

$$VRR = \frac{\dfrac{q_{peak,f} \times (B+h/2)(l/3+h/2)}{0.5 \times P_s} - 1}{\dfrac{q_{peak,b} \times (B+h/2)(l/3+h/2)}{0.5 \times P_s} - 1} \tag{6.13}$$

Using $B = 0.25$ m, $L = 2.4$ m and $h = 0.3$ m, maximum permissible speed and VRR are plotted against VCI at different effective confining pressures for 25 t and 30 t (i.e. static wheel loads of 122.5 kN and 147.2 kN), as shown in Figure 6.7 and Figure 6.8, respectively.

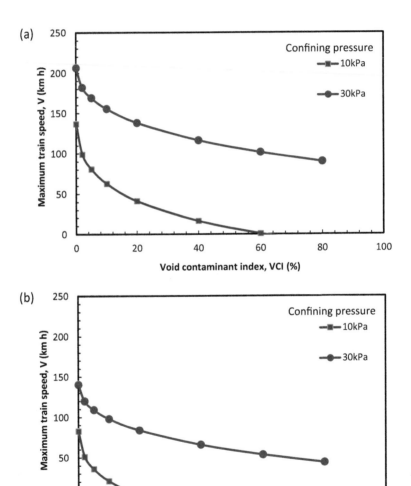

Figure 6.7 Variation of maximum allowable train speed with VCI at different effective confining pressures for (a) P_s = 122.5 kN and (b) P_s = 147.2 kN

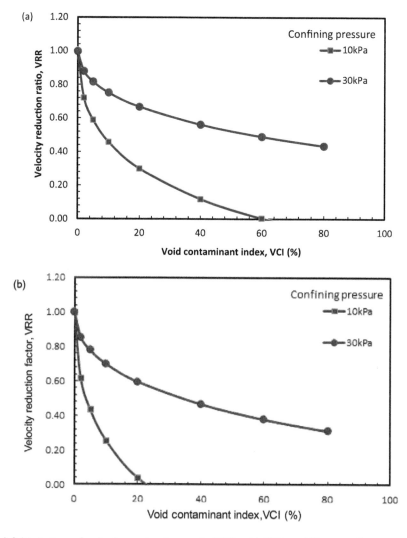

Figure 6.8 Variation of velocity reduction ratio, VRR with VCI at different effective confining pressures for (a) $P_s = 122.5$ kN and (b) $P_s = 147.2$ kN

NOTE: in Figures 6.7a and b, at VCI = 0, the potential speed > 150 km/hour can be too high in a practical sense for heavy haul. They are theoretical maximums and should not be taken as design values for new tracks even if VCI = 0.

6.6 Determining VCI in the field

The method for determining the *in situ* ballast density inspired by Selig and Waters (1994) was used to determine VCI. The ballast was excavated in several layers so that the layers of fouled ballast can be identified properly. Figure 6.9 presents the field test set up. The test device consists of (i) top and bottom cylindrical moulds with known volumes, (ii) the base plate, (iii) the top plate, and (iv) the displacement gauge.

The stepwise procedure is illustrated below where there are two layers (Fig. 6.9):

Step 1: Remove the first layer of ballast and mark (or measure) its thickness to establish a datum, and then fill the hole with a known volume of water.
Step 2: Remove the second layer of ballast.
Step 3: Fill the remaining hole with a known volume of water.
Step 4: Use a 9.5 mm sieve to separate the fouling material from the ballast particles.
Step 5: Determine the dry weights of the clean ballast (M_{b1}, M_{b2}) and the dry weights of the fouling material (M_{f1} and M_{f2}) for layers 1 and 2 respectively.
Step 6: Determine the specific gravities of the ballast particle (G_s) and fouling material ($G_{s,f}$)
Step 7: Calculate the initial void ratio of ballast (e_b) for the initial density of the ballast (ρ_b) when the track was constructed:

$$e_b = \left(\frac{G_{sb}\rho_w}{\rho_b} \right) - 1 \tag{6.15}$$

Step 8: Calculate the void ratio of fouling materials (e_{f1}, e_{f2}) for layers 1 and 2, respectively:

$$e_{f1} = \left(e_b \frac{M_{b1}}{M_{f1}} \frac{G_{sf1}}{G_{sb}} \right) - 1 \tag{6.16}$$

$$e_{f2} = \left(e_b \frac{M_{b2}}{M_{f2}} \frac{G_{sf2}}{G_{sb}} \right) - 1 \tag{6.17}$$

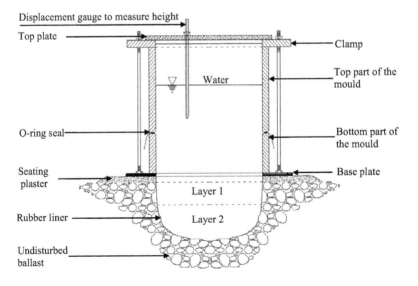

Figure 6.9 Field test set up for determining VCI

Step 9: Determine the VCI for each layer, substituting G_{sb}, G_{sf1}, G_{sf2}, M_{f1}, M_{f2}, M_{b1}, M_{b2}, e_{f1}, e_{f2} and e_b in the Equations 6.18 and 6.19.

$$VCI_1 = \frac{\left(1+e_{f1}\right)}{e_b} \times \frac{G_{sb}}{G_{sf1}} \times \frac{M_{f1}}{M_{b1}} \times 100 \tag{6.18}$$

$$VCI_2 = \frac{\left(1+e_{f2}\right)}{e_b} \times \frac{G_{sb}}{G_{sf2}} \times \frac{M_{f2}}{M_{b2}} \times 100 \tag{6.19}$$

One of the salient benefits of this approach is that it accurately assesses how the fouling materials are distributed within the pore structure of the ballast, a feature that was lacking in previously established indices such as FI and PVC. The track drainage capacity is also governed by the location and extent of fouling, and this information can be obtained accurately by using the field procedure described here. Also, when there are different fouling materials with different values of specific gravities, the resulting different volumes of fouling materials occupying the ballast voids can be captured correctly using VCI, as shown earlier in Figure 6.2.

Chapter 7

Application of geosynthetics in railway tracks

Geosynthetics is the collective term applied to thin, flexible sheets manufactured from synthetic materials (e.g. polyethylene, polypropylene, polyester etc.), which are used in conjunction with soils and aggregates to enhance soil properties (e.g. shear strength, hydraulic conductivity, filtration, separation etc.). Over the past few decades, various types of geosynthetics have been tried out in track to minimise settlement and enhance drainage, but mainly as trial and error exercises (Bergado *et al.* 1993; Han and Bhandari 2009; Indraratna *et al.* 2011a; McDowell *et al.* 2006; Ngo *et al.* 2016; Qian *et al.* 2010; Tang *et al.* 2008, among others). In this chapter, different types of geosynthetics available for geotechnical applications and their effectiveness in harsh railway environment are discussed.

7.1 Types and functions of geosynthetics

Geosynthetics may be classified into two major groups: (i) geotextiles and (ii) geomembranes Geotextiles primarily constitute textile fabrics which are permeable to fluids (water and gas). There are other products closely allied to geotextiles such as geogrids, geomeshes, geonets and geomats, which have all been used in specialised earthfill practices. Unlike most geosynthetics, geomembranes are usually impermeable to water and are mainly used for retention purposes. Different types of geosynthetics have been used in track according to their functions, cost and the engineering properties of the substructure materials, as summarised in Table 7.1 (adopted from Fluet 1986). Geosynthetics can reduce vertical track deformation by controlling lateral movement (through transferring lateral loads from ballast to geosynthetics by shear), dissipate excess pore water pressures developed under repeated loading and protect the ballast from fouling through separation and filtering functions.

7.2 Geogrid reinforcement mechanism

For ballasted railway tracks in particular, geogrids are generally used for reinforcement, which is provided by the tensile strains developed in the geogrids and the interlocking effect between the geogrids and surrounding particles of ballast. There are several types of geogrids, depending on the manufacturing process and how the longitudinal and transverse elements are joined together (Shukla and Yin 2006). The reinforcement mechanism between geogrid and railway ballast is governed by the interlocking effect attributed to the ballast particles partially penetrating through the apertures in the geogrid (Coleman 1990; McDowell and Stickley 2006). Having incorporated geogrids within a ballast layer or at the interface of the ballast and sub-ballast, geogrids interact with the surrounding particles to carry the tensile loads induced by rail vehicles. Through the interlocking and shear interaction between

Table 7.1 Functions and properties of geosynthetics used for rail tracks

Classical function	Railroad function	Relevant geosynthetic properties
Transmission	Transmits water from precipitation and/or pumping through plane of geosynthetic to edge of track	Transmissivity (the product of lateral permeability and thickness)
Filtration	Allows passage of water pumped from subgrade while retaining fines in subgrade	Permeability Retention characteristics Clogging resistance
Separation	Acts as a barrier and prevents intermixing of ballast and sub-ballast/subgrade	Retention characteristics Resistance to concentrated stresses (i.e. tear, puncture, burst)
Reinforcement	Reinforced ballast, may reinforce subgrade and track	Geosynthetic soil/ballast interaction Tensile modulus Tensile strength

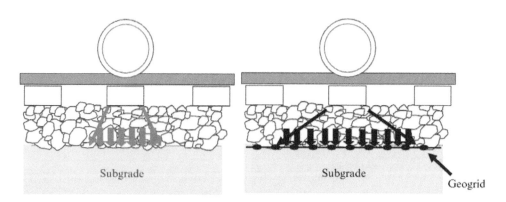

Figure 7.1 Load distribution with and without geogrid reinforcement

the ballast and geogrids, the ballast particles are restrained laterally and tensile forces are transmitted from the ballast to the geogrids. The vertical load applied through the ballast aggregates above the geogrid can now generate tensile resistance in the ribs with very small deflection. Since geogrids are much stiffer in tension than ballast, the lateral stress in the geogrid-reinforced ballast is decreased and reduced settlement can be observed. This inter-action between the geogrid and ballast aggregates increases the shear strength (Indraratna *et al.* 2011b; Ngo *et al.* 2014) and thereby increases the load distribution capacity of track substructure, as reported by Kwon and Penman (2009) and shown in Figure 7.1.

7.3 Use of geosynthetics in tracks – UOW field measurements and laboratory tests

7.3.1 Track construction at Bulli

Geosynthetics have been successfully used in new rail tracks and in track rehabilitation schemes for almost three decades, and when appropriately designed and installed, they are a cost-effective alternative to more traditional techniques (Indraratna *et al.* 2016; Kwon and

Penman 2009). To investigate the stress and deformation imparted to track by train traffic, as well as the benefits of using geosynthetics in fresh and recycled ballast, a field trial has been carried out in a section at Bulli track (Fig. 7.2), owned and operated by Sydney Trains (formerly RailCorp). During this period, the train-induced stresses and the vertical and lateral deformations of the track were monitored by the Centre for Geomechanics and Railway Engineering, University of Wollongong (Indraratna *et al.* 2010).

The construction and instrumentation of this track segment include subgrade consisting of stiff, over-consolidated silty clay with shale cobbles and gravels over bedrock of highly weathered sandstone; the layers of ballast and sub-ballast are 300 mm and 150 mm thick, respectively. There are four track sections built; fresh and recycled ballast without inclusion of a geocomposite are used at Sections 1 and 4, whereas the other two sections are reinforced by a layer of geocomposite at the ballast–sub-ballast interface (details of track construction described by Indraratna *et al.* 2010). The geocomposite is composed of a biaxial geogrid (aperture size = 40 mm × 27 mm, peak tensile strength = 30 kN m^{-1}) placed over a layer of non-woven polypropylene geotextile (mass per unit area = 140 g m^{-2}, thickness = 2 mm). Further technical specifications of the materials used during construction are reported elsewhere by Indraratna *et al.* (2016).

(a) placement of biaxial geogrid at Bulli tracks

(b) installation of strain gauges on geogrid

(c) installation of settlement pegs

(d) displacement transducers

Figure 7.2 Installation of geogrids at Bulli tracks

The vertical and horizontal stresses are measured by rapid response hydraulic earth pressure cells with thick, grooved active faces based on semi-conductor type transducers. Settlement pegs were installed between the sleeper and ballast, and between the ballast and sub-ballast to measure the vertical deformation of the ballast layer. These settlement pegs consist of 100 mm × 100 mm × 6 mm stainless steel base plates attached to 10 mm diameter steel rods. Lateral deformation could be recorded by potentiometric displacement transducers placed inside 2.5 m long stainless-steel tubes that could slide over each other, with 100 mm × 100 mm end caps as anchors. The pressure cells and lateral displacement transducers were connected to a computer-controlled data acquisition system which could operate at a maximum frequency of 40 Hz.

7.4 Measured ballast deformation

In the field, vertical and horizontal deformation is measured against time, which means that a relationship between the annual rail traffic in million gross tonnes (MGT) and axle load (A_t) is needed to determine the number of load cycles N, as proposed by Selig and Waters (1994). This relationship is expressed as: $N_t = 106/(A_t \times N_c)$, where N_t, A_t and N_c are the numbers of load cycles per MGT, the axle load in tonnes and the number of axles per load cycle. When this relationship is used for a traffic tonnage of 60 MGT per year and four axles per load cycle, an axle load of 25 tonnes gives 600,000 load cycles per MGT. A simple survey technique is then used to record changes in the reduced level of tip of the settlement peg (Indraratna et al. 2010). Figure 7.3 shows the variation of average deformation of ballast against the number of load cycles (N). Unlike fresh ballast, recycled ballast exhibits less vertical and lateral

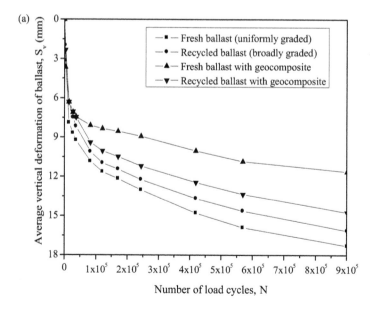

Figure 7.3 Average deformation of the ballast layer: (a) vertical settlement (S_v); (b) lateral displacement (S_h)

(data sourced from Indraratna et al. 2010 – with permission from ASCE)

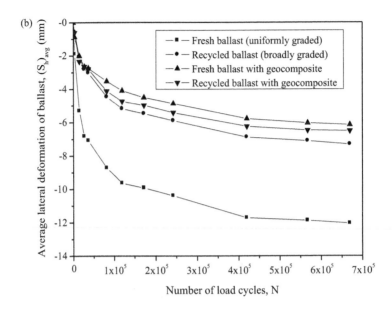

Figure 7.3 (Continued)

deformation, possibly due to its moderately graded particle size distribution – PSD ($C_u = 1.8$) compared to the very uniform PSD ($C_u = 1.5$) of fresh ballast. These results also indicate that the geocomposite reinforcement reduced the vertical (S_v) and lateral (S_h) deformation of fresh ballast by about 33% and 49%, respectively, while decreasing the vertical and lateral deformation of recycled ballast by about 9% and 11%, respectively. Lateral deformation is one of the most important indices affecting track stability, and the use of a geocomposite layer can be an effective way of curtailing it significantly, with obvious implications for improved track performance and reduced maintenance costs.

7.5 Traffic-induced stresses

Figure 7.4a shows the peak cyclic vertical (σ_v) and lateral (σ_l) stresses recorded at Section 1 (i.e. fresh ballast without geocomposite) after the passage of a coal train with an axle load of 25 tonnes. Here, the peak cyclic vertical stress decreased by 73% and 82% at depths of 300 mm and 450 mm, respectively. Moreover, σ_l decreased only marginally with depth, which implies that artificial inclusions are needed for additional restraints (Nimbalkar *et al.* 2012). While most of the peak cyclic vertical stresses were below 230 kPa, one value of σ_v reached 415 kPa, as shown in Figure 7.4b; this was later found to be associated with a wheel flat, thus proving that much larger stresses are induced by wheel imperfections (Kaewunruen and Remennikov 2010).

7.6 Optimum geogrid size for a given ballast

A series of large-scale direct shear tests for average-sized ballast particles (d_{50}) of 35 mm, reinforced by seven types of geogrids with apertures varying from 20.8 mm to 80 mm, were

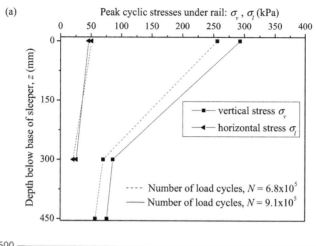

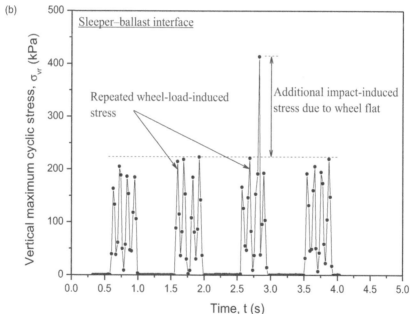

Figure 7.4 Cyclic stresses induced by coal train with wagons (100 tonnes): (a) variation of stresses with depth; (b) additional stress due to wheel flat

(data sourced from Indraratna *et al.* 2010 – with permission from ASCE)

conducted to investigate how the size of the aperture affects the shear behaviour at the bal-last–geosynthetics interfaces. The laboratory results indicated that the shear strength at the interface was governed by the size of the geogrid aperture. They adopted an interface effi-ciency factor (α), defined by Koerner (1998), to evaluate the improvement of shear behaviour at the geogrid-reinforced ballast interface, where the efficiency factor (α) at the interface is given by:

$$\alpha = \frac{\tan(\delta)}{\tan(\phi)} \tag{7.1}$$

where δ is the apparent friction angle of the interface and ϕ is the friction angle of the soil. The interface efficiency factors (α) of ballast reinforced by different geogrids and plotted with ratios (A_g/d_{50}) of geogrid aperture size (A_g) to the average size particle of ballast (d_{50}) are shown in Figure 7.5. Here the value of α depends on the ratio of A_g/d_{50} until it reaches a maximum value of 1.16 at A_g/d_{50} of 1.21, followed by a gradual decrease towards unity as the ratio of A_g/d_{50} reaches 2.5. Furthermore, the value of α is less than unity (i.e., unreinforced ballast specimen) for the ratio of $A_g/d_{50} < 0.95$, and when $A_g/d_{50} > 0.95$, the value of α is greater than unity. This observation can be attributed to the interlocking between geogrids and ballast aggregates, where $\alpha < 1$ shows an ineffective interlocking and $\alpha > 1$ represents an acceptable interlocking. Based on the data presented in Figure 7.5, Indraratna *et al.* (2012a) classified the ratio of A_g/d_{50} into three main zones:

- Feeble interlock zone (FIZ): when $A_g/d_{50} < 0.95$ the ballast–geogrid interlock is weaker than the aggregate-to-aggregate interlock observed without geogrid. This is because the ballast–geogrid interlock is mainly attributed to smaller aggregates alone ($<0.95\ d_{50}$) when compared to the aggregate-to-aggregate interlock considered for all sizes. In this zone they observed an insignificant ballast breakage after tests, suggesting that failure at the interface is caused by a loss of the ballast–geogrid interlock during shearing.
- Optimum interlock zone (OIZ): when A_g/d_{50} varied from 0.95 to 1.20, the interlocking of larger ballast aggregates with geogrid occurs, resulting in an increased value of α until it reaches a maximum value of 1.16 at an optimum A_g/d_{50} ratio of 1.20. After testing, there was a lot of broken ballast at the interface, which indicated that failure at the interface stemmed from the breakage of initially interlocked ballast.
- Diminishing interlock zone (DIZ): when $A_g/d_{50} > 1.2$, the value of α in this zone is greater than unity, but then decreases with an increase in the ratio A_g/d_{50}, showing reduced interlocking. It was indicated that the reduction of α could be attributed to the amount of ballast trapped within a given aperture which can displace itself freely, as the unreinforced case.

The interface efficiency depicted in Figure 7.5 is based on laboratory tests conducted on fresh latite ballast–geosynthetic interfaces carried out using the large-scale direct shear apparatus. Fresh latite ballast conforming to the standards specified by technical specification TS 3402 and a particle size distribution (PSD) that conformed to AS 2758.7 were adopted. It is noted that these results are based on testing conditions carried out at the UOW, including: type of geogrid to be used for reinforcement is biaxial geogrid; geogrid tensile strength is generally between 15 and 35 kN m^{-1}; and the placement location of geogrid is at the sub-ballast–ballast interface.

7.7 Role of geosynthetics on track settlement

The response of fresh and recycled ballast under cyclic loading was investigated in a laboratory model apparatus in both dry and wet states. Figures 7.6a and b show the settlements of

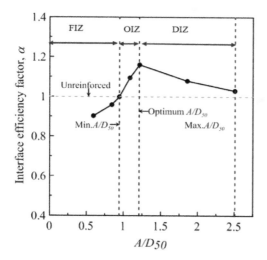

Figure 7.5 Interface efficiency factor (α) versus A_g/d_{50}, a dimensionless parameter (modified after Indraratna *et al.* 2012a)

fresh and recycled ballast, both dry and wet, and with and without inclusion of geosynthetics. As expected, fresh dry ballast gives the least settlement. It is believed that the relatively higher angularity of fresh ballast contributes to better particle interlock and therefore gives less settlement. Recycled ballast without any geosynthetic inclusion exhibits significantly higher settlement compared to fresh ballast, especially when they are wet (saturated). The reason for this is that reduced angularity of recycled ballast results in less friction angle and lower deformation modulus compared to fresh ballast. The test results show that wet recycled ballast (without any geosynthetic inclusion) gives the highest settlement, because water acts as a lubricant, which reduces frictional resistance.

Figure 7.6 shows the benefits of using geosynthetics in recycled ballast (both dry and wet). Each of the three types of geosynthetics used in this study decreases the settlement considerably. However, the geocomposite (geogrid bonded with a non-woven geotextile) stabilises recycled ballast remarkably well. The combination of reinforcement by the geogrid and the filtration and separation provided by the non-woven geotextile (of the geocomposite) minimises the lateral spreading and fouling of recycled ballast, especially when wet. The non-woven geotextile also prevents the fines moving up from the capping and subgrade layers and keeps the recycled ballast relatively clean. Figure 7.6 shows one common feature; initially the settlement increases rapidly in all specimens. It was also noted that all ballast specimens stabilised within about 100,000 load cycles, beyond which the settlement increase was marginal.

7.7.1 Predicted settlement of fresh ballast

The extensive large-scale laboratory tests carried out at the UOW indicated that ballast settlement under cyclic loading can be represented by the following semi-logarithmic relationship:

$$Settlement(mm) = a + b \times ln(N) \tag{7.2}$$

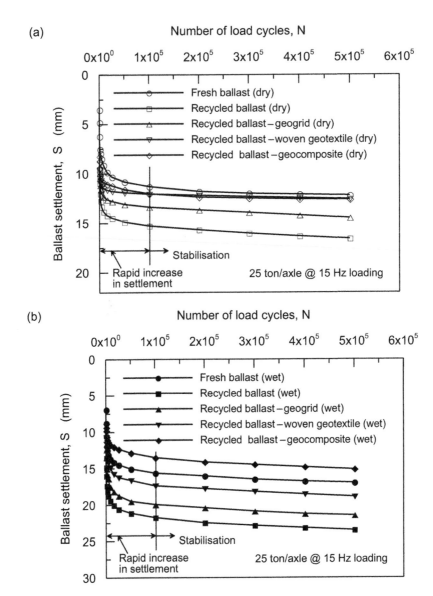

Figure 7.6 Settlement response of fresh and recycled ballast under cyclic loading, (a) in dry condition and (b) in wet condition

(data source: Salim 2004 – PhD thesis)

where N is the number of load cycles and a and b are two empirical constants, depending on the type of ballast and type of geosynthetics used, and given in Table 7.2. Also, loading conditions conducted were: confining pressure: 10 kPa; vertical stress: 60 kPa; and frequency: 15 Hz.

Table 7.2 Values of empirical coefficients *a* and *b* in case of fresh ballast

Reinforcement type	a	b
Unreinforced	2.94	0.56
Geogrid	2.09	0.48
Geotextile	3.72	0.39
Geocomposite	1.51	0.50

7.7.2 Predicted settlement of recycled ballast

The vertical settlement of recycled ballast with and without reinforcement could be given by the following equation:

$$Settlement(mm) = c + d \times ln(N) \tag{7.3}$$

where c and d are the empirical constants and N is the number of cycles. The values of c and d in case of unreinforced (recycled) ballast and geogrid-, geotextile- and geocomposite-reinforced ballast are summarised in Table 7.3.

7.7.3 Settlement reduction factor

The reduction in settlement could be represented by means of a "settlement reduction factor (SRF)", which can be defined as the ratio between the difference in the settlement of reinforced and unreinforced ballast, given by:

$$SRF = 1 - \frac{Settlement_{reinforced}}{Settlement_{unreinforced}} \tag{7.4}$$

where SRF is the settlement reduction factor and $Settlement_{reinforced}$ and $Settlement_{unreinforced}$ are the settlements of reinforced and unreinforced ballast.

7.7.4 The effect of fouling on the ballast–geogrid interface shear strength

Fouling is caused by fine particles that accumulate in the voids of ballast. Therefore, fouled ballast should ideally be simulated in DEM by injecting various amounts of fine particles into the voids to represent different values of VCI (Fig. 7.7a). Owing to fouling material between the individual rough and angular particles of ballast, the inter-particle friction angle is expected to decrease (Fig. 7.7b). This reduction in the apparent angle of friction is evaluated experimentally and presented in the following sections.

In this section, the effect of fouling on the ballast-geogrid interface shear strength is presented. This gives the practitioner an idea of how the friction angle and interface shear strength at the ballast–geogrid interface decreases due to the effect of fouling. Large-scale direct shear tests (Fig. 7.8) for fresh and coal-fouled ballast reinforced by a 40 mm × 40 mm geogrid were carried out to a maximum horizontal displacement of $\Delta h = 37$mm, under different normal stresses of $\sigma_n = 15, 27, 51$ and 75kPa. During the shearing process, the shearing

Table 7.3 Values of empirical coefficients c and d in case of recycled ballast

Reinforcement type	c	d
Unreinforced	9.12	0.53
Geogrid	7.71	0.50
Geotextile	6.87	0.46
Geocomposite	6.45	0.48

Table 7.4 Settlement reduction factor for fresh and recycled ballast

Reinforcement type	SRF (N = 500,000)	
	Fresh ballast	Recycled ballast
Geogrid	0.18	0.11
Geotextile	0.14	0.19
Geocomposite	0.21	0.21

forces and vertical displacements of the top plate were recorded at every 1 mm of horizontal displacement. The shear stresses and vertical strains were then computed and plotted against the horizontal shear strain.

Designers are required to enter the degree of fouling in terms of the parameter VCI and the applied normal stress on ballast (i.e. confining pressure, σ_n). UOW test data currently incorporate the data corresponding to the applied normal stress of 15, 27, 51 and 75 kPa only based on the study carried out by Indraratna et al. (2011a), and hence the appropriate value of the normal stress should be selected by the users. Similarly, the value of the VCI should be any of the values indicated here in brackets (0, 10, 20, 40, 70 and 95). An appropriate value of the applied normal stress and the VCI can be selected subject to further tests that are appropriate for design and testing conditions. The friction angle of the unreinforced ballast specimen and of that reinforced with a geogrid of aperture 40 × 40 mm is given for different levels of coal fouling.

Laboratory test results indicate that the peak shear stress of ballast increases with normal stress and decreases with an increasing level of fouling. Strain softening and dilation are also observed in all the tests, where a higher normal stress σ_n results in a greater shear strength and in smaller dilations. The coal fines can reduce the peak shear stresses of the reinforced and unreinforced ballast assemblies because they coat the surfaces of ballast grains, thus inhibiting inter-particle friction and reducing the shearing resistance at the geogrid–ballast interface. The variations of normalised peak shear stress (τ_p/σ_n) and the apparent angle of shearing resistance (ϕ) with VCI for fouled ballast assemblies with and without geogrid reinforcement are shown in Figure 7.9. Note that coal fines steadily reduce the peak shear stress of a ballast assembly, which then diminishes the apparent angle of shearing resistance. This reduction of (τ_p/σ_n) due to the presence of coal fines is significant when the VCI is less than 70%, but it becomes marginal when the VCI is higher.

The effect of fouling materials on the reduced shear strength is illustrated in Figure 7.10. The normalised reduction in shear strength is expressed as the ratio between the decrease in peak shear stress ($\Delta\tau_p$) and normal stress (σ_n). Figure 7.10 shows that the decrease in shear strength is greater for unreinforced ballast than for ballast stabilised by geogrid, and this is

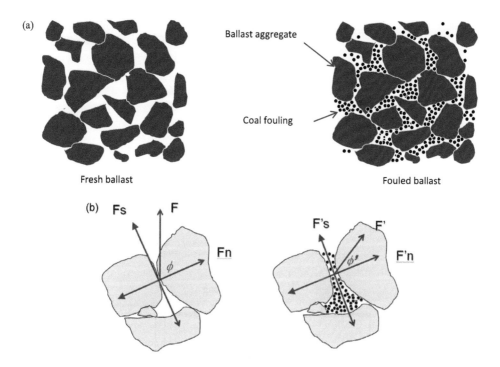

Figure 7.7 (a) Conceptual sketch of fouled ballast; (b) decreased friction angle of ballast due to fouling

(modified after Indraratna *et al.* 2014 – with permission from ASCE)

believed to be due to the interlocking effect created at the ballast–geogrid interface (Qian *et al.* 2010; Raymond 2002). The variations of the decrease in normalised peak shear stress for ballast with and without geogrid, with respect to changes in the *VCI*, could be described by the following hyperbolic equation:

$$\frac{\Delta\tau_p}{\sigma_n} = \frac{VCI/100}{a \times VCI/100 + b} \tag{7.5}$$

where $\Delta\tau_p$ = shear strength reduction of ballast due to the presence of fines; σ_n = normal stress; VCI = void contamination index; and a and b = hyperbolic constants.

The results obtained from direct shear tests on ballast with and without geogrid reinforcement are plotted in transformed axes to determine the hyperbolic constants (a, b) by rearranging Equation 2 as follows:

$$\frac{VCI}{100} \times \frac{\sigma_n}{\Delta\tau_p} = a \times \frac{VCI}{100} + b \tag{7.6}$$

The linear regression curves presented in Figure 7.11 prove that a decrease in normalised peak shear stress could be estimated accurately based on a hyperbolic relationship (coefficient of

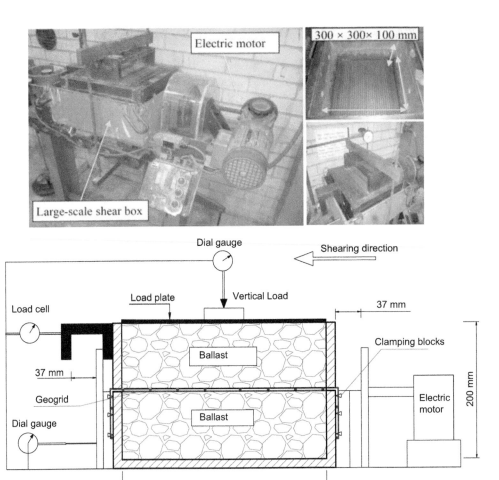

Figure 7.8 Large-scale direct shear apparatus used in the laboratory

regression, $R^2 > 0.95$). The hyperbolic constants *a* and *b* are, for both cases, presented in tabular forms in Figure 7.11. It is observed that *a* and *b* are independent of the *VCI* ratio (fines content) and vary with applied normal stresses.

7.8 The effect of coal fouling on the load-deformation of geogrid-reinforced ballast

7.8.1 Laboratory study using process simulation testing apparatus

To investigate the effects that geogrid and coal fines have on the deformation and degradation of fresh and fouled ballast, a series of tests using the process simulation testing apparatus (PSTA) was carried out to simulate realistic track conditions. The novel PSTA was modified

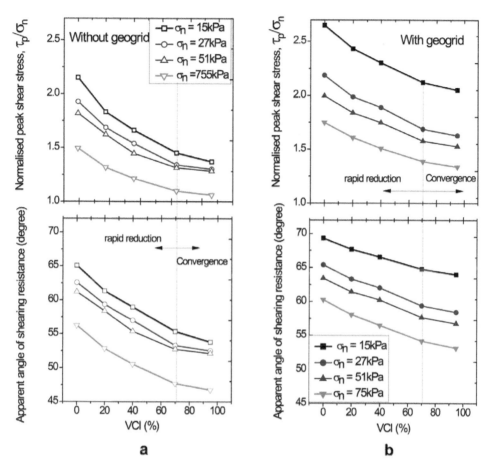

Figure 7.9 Effect of *VCI* on the normalised peak shear strength and apparent angle of shearing resistance of ballast: (a) without geogrid; (b) with geogrid

(after Indraratna et al. 2011a)

based on an original design by Indraratna and Salim (2003), as illustrated in Figure 7.12. Typically, for tested ballast gradations in NSW, the initial stresses were kept constant around 6–7 kPa in the transverse direction (parallel to sleeper), and about 10–12 kPa along the longitudinal direction, for which the lateral strains must be kept as small as possible (i.e. plane strain). A lateral stress ratio of 0.5–0.7 was typical for maintaining the plane strain for conventional axle loads. The lateral pressures were selected based on the lateral confinement provided by the weight of crib and shoulder ballast, along with particle frictional interlock. Similar lateral pressures (and stress ratios) were used in previous studies, as reported by Indraratna and Salim (2003) based on actual track measurements. Generally, the stress applied in the laboratory is controlled to vary within small limits in the longitudinal direction to ensure insignificant strains (perpendicular to sleeper). The lateral stress ratio adopted

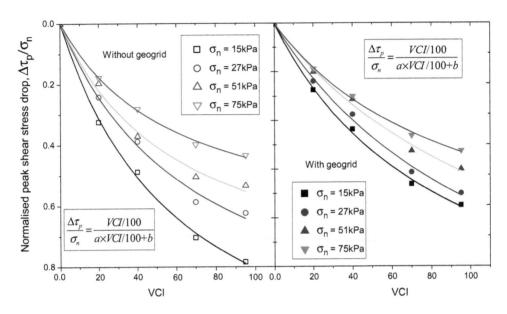

Figure 7.10 Variation of normalised peak shear stress drop for unreinforced and biaxial reinforced-ballast with *VCI*

(data source from Indraratna et al. 2011a)

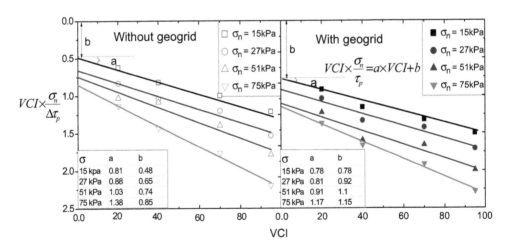

Figure 7.11 Determination of hyperbolic constants a and b for ballast with and without geogrid reinforcement

(data source from Indraratna et al. 2011a)

in this study is justified by field measurements from Singleton and Bulli tracks in NSW, Australia.

The PSTA used in this study consists of four main components: the prismoidal triaxial chamber, the axial loading unit, the confining pressure control system, and the horizontal and vertical displacement monitoring devices. A schematic cross-section and plan view of the PSTA are illustrated in Figures 7.12a and b. A general view and typical parts of the PSTA is shown in Figure 7.12c. The PSTA can accommodate a ballast assembly of 800 mm long by 600 mm wide by 600 mm high. All four vertical walls of the PSTA are placed inside the frame and supported on the displacement system. There is 1 mm gap between the four vertical walls and the base plate, which allows free movement of the vertical walls when subjected to a horizontal force. A system of hinges and ball bearings are lubricated regularly to minimise frictional resistance and enable the vertical walls to displace laterally with minimum friction. Eight steel pegs are placed at each of the sleeper–ballast and ballast–sub-ballast interfaces to measure vertical settlement and help calculate the vertical strain of the ballast layer.

A cyclic load is applied by a servo hydraulic actuator and transmitted through the ballast by a wooden sleeper connected to a steel rail. A linear variable differential transformer

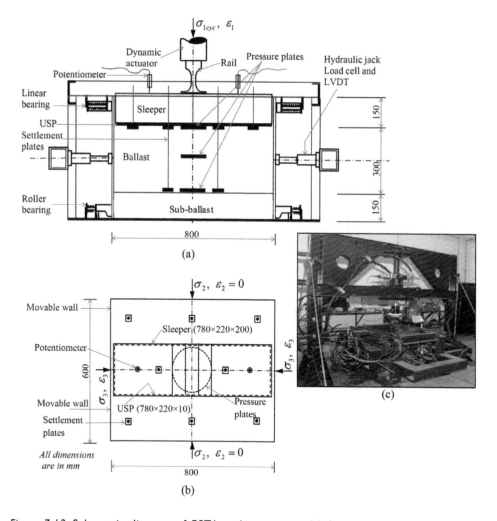

Figure 7.12 Schematic diagram of PSTA and test setup: (a) front view; (b) plan view; and (c) a photograph of triaxial cubical chamber of PSTA

(photo taken at SMART-Rail lab)

(LVDT) is connected to the load actuator to record its vertical movement. Confining pressures are applied in two horizontal directions (perpendicular and parallel to the sleeper) by hydraulic jacks connected with load cells to control the pressure applied during testing. Lateral movements of the four vertical walls were measured with 16 electronic potentiometers (Fig. 7.12b).

The vertical stresses at the sleeper–ballast and ballast–sub-ballast interfaces are measured by two pressure plates. A typical harmonic cyclic load applied in this study is estimated in accordance with Esveld (2001) and presented in Figure 7.13. All the tests are conducted at a frequency of 15 Hz with a maximum induced cyclic pressure of 420 kPa and are tested up to 500,000 load cycles. A frequency of 15 Hz was selected based on the freight lines operating at approximately 100 km h. In Australia, the axle loads vary from 25 to 30 tonnes. In the field, a wheel load is transmitted vertically (underneath sleeper) and laterally to adjacent sleepers. Atalar et al. (2001) reported that part of the wheel load is transmitted to the adjacent sleepers, and only 40–60% of the wheel load is actually carried by the sleeper beneath the wheel. Therefore, in the test setup, an applied load of 20 tonnes over a single concrete sleeper (650 mm long and 220 mm wide) is expected to generate a stress of 550–800 kPa at the sleeper–ballast interface.

The author's current experimental measurements in the test rig are around 420 kPa, and some are even smaller. The field measurements made in several Australian case studies, including the Bulli and Singleton tracks, have shown that stresses just beneath the sleeper are from 350–500 kPa (Indraratna et al. 2011b). Therefore, the 20-tonne load applied in the experimental rig is justified because it generates a realistic stress on the ballast. Every instrument is calibrated before being connected to an electronic data logger DT800 (Fig. 7.14a) and then is controlled by a host computer, supported by Labview software, to accurately record the vertical settlement, pressures, and the lateral displacement of associated walls at predetermined time intervals during the testing phase. The data acquisition screen is shown in Figure 7.14b.

7.8.2 Materials tested

Samples of fresh latite ballast are collected from Bombo quarry, New South Wales, Australia, then cleaned and sieved according to Australia Standards (AS 2758.7 1996). The size and characteristics of ballast and sub-ballast used in this study are shown in Table 7.5. The particle size distribution curve (PSD) of ballast and sub-ballast used in this study is shown in Figure 7.15. Coal fines similar to those used for the large-scale direct shear tests were also

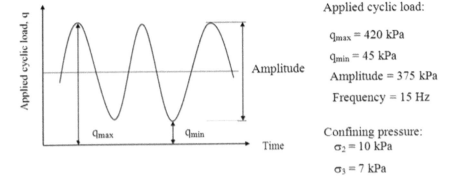

Applied cyclic load:

q_{max} = 420 kPa

q_{min} = 45 kPa

Amplitude = 375 kPa

Frequency = 15 Hz

Confining pressure:

σ_2 = 10 kPa

σ_3 = 7 kPa

Figure 7.13 Typical cyclic loading applied in the study

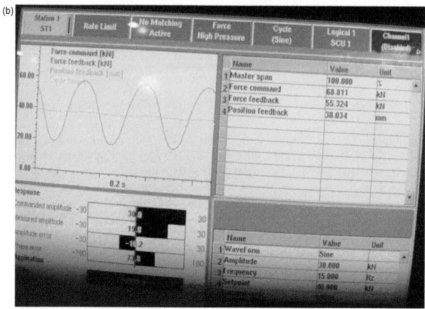

Figure 7.14 (a) Data logger DT800 to record displacement during testing; (b) data acquisition screen

Table 7.5 Grain size characteristics of ballast and sub-ballast used for the PSTA tests

Test type	Particle shape	d_{max} (mm)	d_{10} (mm)	d_{30} (mm)	d_{50} (mm)	d_{60} (mm)	C_u	C_c	Size ratio
Ballast	Highly angular	53	16	28	35	39	2.4	1.3	11.3
Sub-ballast	Angular to rounded	19	0.23	0.45	0.61	0.8	3.5	1.1	31.6

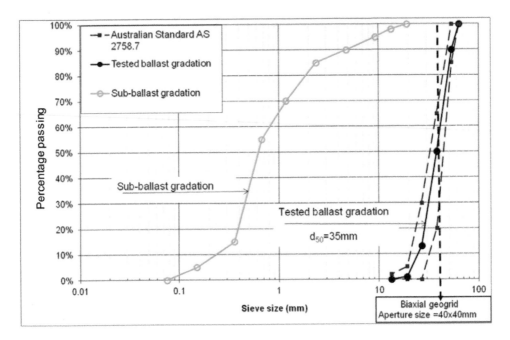

Figure 7.15 Particle size distribution of ballast and sub-ballast used in the PSTA tests

used for all the PSTA tests. The coal fines are provided by Queensland Rail and used as fouling material for VCIs of 10%, 20%, 40% and 70%. The engineering characteristics of coal fines are presented in Table 7.6. Given the prolonged droughts that occur in Queensland and the hot climate that prevails almost year round in most states in Australia, the coal fouled tracks are often dry and unsaturated, as one may also find in some North American tracks, for instance as tested by Tutumluer *et al.* (2008). Therefore, the coal fines and ballast tested in this study were relatively dry (moisture content is less than 4%). It is agreed that the plastic properties of coal dust can be affected if mixed with plastic clay in the field, but in Australia, the vast majority of coal lines are mainly fouled by coal falling off the wagons during the passage of coal freight trains, so there is almost no mixing with clay.

Apart from carbon–hydrogen–nitrogen studies, various mineral beneficiation methods such as iron chromatography, atomic absorption and X-ray diffraction tests carried out in the past have shown that fouled coal from Queensland tracks contains less than 6–8% of clay minerals, so they add no plasticity to coal fouling. It is also true that *VCI* in the field can vary with time due to ballast degradation and the rate of infiltration of coal fines, and that the fouling rate can also vary with location and time. In this study, the amount of coal fines is kept constant for each test to enable proper comparisons. It is expected that track inspection

Table 7.6 Engineering properties of coal fines tested

	Specific gravity	Liquid limit (%)	Plastic limit (%)	Optimum moisture content, OMC (%)	Maximum dry density (kg m^{-3})	Passing no. 200 sieve (%)	Mean particle size d_{50} (mm)
Coal fine	1.28	71	41	35	874	15	1.18

would be mandatory to ensure realistic evaluation of track performance. Biaxial geogrid manufactured from polypropylene, with 40 mm × 40 mm square apertures similar to the geogrid tested in the large-scale direct shear tests, is used in this study. To ease the recovery of ballast after each test, a non-woven 2 mm thick geotextile is placed above the layer of sub-ballast and beneath the geogrid to act as a separator so that it would not affect the inter-lock between ballast particles and geogrid. This type of geotextile is commonly used in rail tracks in Australia to separate the layers of ballast and sub-ballast. The geotextile is very thin (2 mm) and is placed loosely without tension so there is almost no reinforcement effect.

7.8.3 Cyclic testing program

A total of ten tests are conducted with and without geogrids, and with *VCI* varying from 0% to 70%. While the samples are being compacted, the side walls of the PSTA are clamped to prevent any deformation. Layers of subgrade and sub-ballast are prepared using a vibratory compactor to reach desired unit weights, as mentioned previously. Pressure plates, settlement pegs and geogrid are then placed onto the layer of sub-ballast. The ballast is divided and compacted into five equal sub-layers (60 mm thick), each of which is compacted with a hand vibrator until the simulated field unit weight (typically at 15.3 kN m^{-3}). A rubber pad is placed beneath the vibrator to prevent particle breakage during compaction. A total of five sub-layers of fresh ballast (60 mm for each sub-layer) are compacted, and then the coal dust is blown through the ballast voids to simulate field conditions where coal fines fall off wagons and infiltrate the ballast bed. This technique also prevents the ballast structure from being disturbed. The subsequent layers of ballast and coal fines are also compacted until the ballast attained its final height of 300 mm. A wooden sleeper is then placed on top of the ballast and connected to a hydraulic actuator via a steel ram. Eight settlement pegs are then placed on top of the ballast, and then more ballast is placed onto the top level of the sleeper to represent crib and shoulder ballast.

After the assembly is prepared, the clamps are removed and lateral pressures (σ_2 =10 kPa and σ_3 = 7 kPa) corresponding to confining pressures typically provided by crib and shoulder ballast on a real track are applied. An initial vertical pressure of 45 kPa is then applied to stabilise the sleeper–ballast assembly and serve as a reference for all lateral displacement and settlement readings. A cyclic load is then applied through a servo hydraulic actuator to a maximum pressure of 420 kPa at a frequency of 15 Hz. These loading characteristics induce an approximately mean contact stress of 233 kPa onto the sleeper and ballast, which represents a 20 tonne/axle train loading travelling at approximately 80 km h under typical Australian track conditions. A total of half a million load cycles are applied in each test, but they are interrupted at specific cycles (1, 10, 100, 1000, 2000, 4000, 7000, 15,000, 30,000, 50,000, 100,000, 200,000, 300,000, 400,000, 500,000) to take readings of settlement pegs and capture the resilience of ballast material at the end of these cycles. The rest periods are captured when selecting these specific cycles and test interruptions are carried out accordingly. Lateral displacements and vertical stresses are automatically recorded by an automated data logger DT800.

7.8.4 Lateral deformation of fresh and fouled ballast

The lateral deformation of fresh and fouled ballast in both horizontal directions and with and without geogrid (perpendicular and parallel to sleeper) are presented in Figures 7.16a and b. It can be seen here that the geogrid reduces the lateral displacement of fresh and fouled ballast by a large amount; in fact, when particles of ballast are compacted over the geogrid, they partially penetrate and project through the apertures and create a strong mechanical interlock between the geogrid and now restrained ballast (Ngo 2012). This interlocking effect enables the geogrid to act as a non-horizontal displacement boundary that confines and restrains the ballast from free movement, which in turn decreases its deformation. This supports previous studies by Konietzky *et al.* (2004) and McDowell *et al.* (2006), where the discrete element method was used to investigate the interaction between geogrid and ballast. They concluded that the geogrid provides an interlocking effect by creating a stiffened zone on each side. During cyclic loading and associated vibrations, ballast particles can rotate and move, so the initial void arrangement and the contact distribution would change accordingly. An increased VCI results in a much larger lateral displacement because when fouling increases, it is highly likely that fouling material (e.g. fines coal dust) clings to the ballast grain surfaces (a process sometimes referred to as armouring) and intrudes between the contact points, thus offering a lubricating effect, which in turn helps the particles of ballast slide and roll over each other. However, the ability of geogrid to reduce lateral displacement also decreases when the VCI increases because when the ballast layer is removed after the test, the coal fines that accumulate in the apertures of the geogrids reduce the geogrid aperture.

7.8.5 Vertical settlements of fresh and fouled ballast

The average accumulated vertical settlement of fresh and fouled ballast at selected load cycles is measured using settlement pegs and a linear variable differential transformer (LVDT). Figure 7.16c shows the settlement of fresh and fouled ballast assemblies compared to geogrid-reinforced ballast at various *VCI*. The settlement of geogrid-reinforced ballast is generally less than the unreinforced assembly for any given *VCI*, although fresh ballast reinforced with geogrid shows the least settlement. As expected, an increasing level of fouling causes more ballast deformation. All the samples have the same initial rapid settlement up to 100,000 cycles, followed by gradually increasing settlement within 300,000 cycles, and then remain relatively stable to the end (500,000 cycles). This clearly indicates that ballast undergoes considerable rearrangement and densification during the initial load cycles, but after attaining a threshold compression, any subsequent loading would resist further settlement and promote dilation when the ballast specimen could not compress any further (i.e. $\varepsilon_2 + \varepsilon_3 > \varepsilon_1$). This increased settlement was also attributed to coal fines acting as a lubricant, as mentioned earlier.

The measured data is best interpreted by Figure 7.17a–d, which plot the final values of deformation and the relative deformation factor at $N = 500,000$ with varying *VCI*. The relative deformation factor (R_f) can be defined as follows:

$$\text{Vertical settlement (\%):} \quad R_s = \frac{S_{(unreinforced)} - S_{(reinforced)}}{S_{(unreinforced)}} \times 100 \tag{7.7}$$

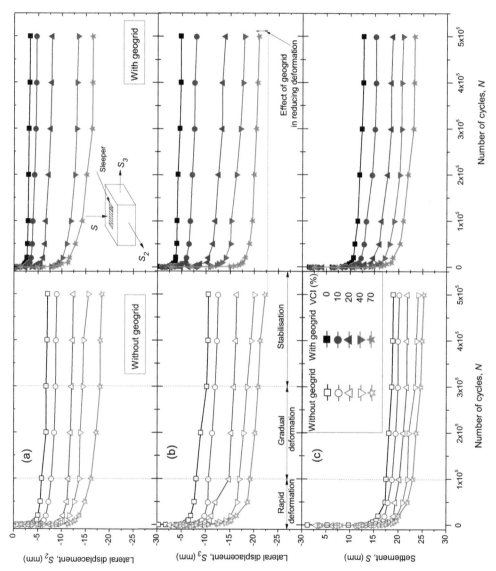

Figure 7.16 Variations of the deformation of fresh and fouled ballast with and without geogrid with varying *VCI*: (a) lateral displacement (S_2, perpendicular to sleeper), (b) lateral displacement (S_3, parallel to sleeper), (c) settlement, S (data source: Indraratna *et al.* 2013a – with permission from ASCE)

$$\text{Lateral deformation } (\%): \quad R_{h2} = \frac{S_{2(unreinforced)} - S_{2(reinforced)}}{S_{2(unreinforced)}} \times 100 \tag{7.8}$$

$$\text{Lateral deformation } (\%): \quad R_{h3} = \frac{S_{3(unreinforced)} - S_{3(reinforced)}}{S_{3(unreinforced)}} \times 100 \tag{7.9}$$

The beneficial effect of geogrid on reducing ballast deformation, as expected, is clearly reflected by the values of R_f presented in Figure 7.17d. The benefit of the geogrid decreases with an increase of *VCI* and becomes marginal when *VCI* > 40%. Geogrid can reduce the deformation of fresh ballast (by approximately 52% and 32% reduction for lateral and vertical deformation, respectively), but this value significantly decreases with an increase of *VCI* (approximately 12% and 5% reduction for lateral and vertical deformation, respectively, for *VCI* = 40%). It is clear that as the *VCI* increases beyond 40%, the subsequent decrease in the value of R_f is gradual compared to fresh and fouled ballast with *VCI* = 10% and 20%. This observation is supported by the fact that at *VCI* = 40% and beyond, coal fines fill the ballast voids and geogrid apertures. This phenomenon inhibits inter-particle friction and prevents the ballast particles from effectively interlocking with the geogrid. Based on this data, it is possible to propose a threshold value of *VCI* = 40% where the effect of geogrid becomes marginal and track maintenance would become imperative.

7.8.6 Average volumetric and shear strain responses

The average vertical strain of the ballast layer $(\varepsilon_1)_{avg}$ is calculated based on the differences between settlement at the sleeper–ballast and ballast–sub-ballast interfaces, as measured by the settlement pegs. The average lateral strains perpendicular and parallel to the sleeper $((\varepsilon_2)_{avg}, (\varepsilon_3)_{avg})$ are then calculated using the lateral displacement of the four vertical walls, as measured by the potentiometers. The average volumetric strain $(\varepsilon_v)_{avg}$ and shear strain $(\varepsilon_s)_{avg}$ can be calculated as follows (Timoshenko and Goodier 1970):

$$(\varepsilon_v)_{avg} = (\varepsilon_1)_{avg} + (\varepsilon_2)_{avg} + (\varepsilon_3)_{avg} \tag{7.10}$$

$$(\varepsilon_s)_{avg} = \frac{\sqrt{2}}{3} \left\{ \sqrt{\left[(\varepsilon_1)_{avg} - (\varepsilon_2)_{avg}\right]^2 + \left[(\varepsilon_2)_{avg} - (\varepsilon_3)_{avg}\right]^2 + \left[(\varepsilon_3)_{avg} - (\varepsilon_1)_{avg}\right]^2} \right\} \tag{7.11}$$

Figure 7.18a shows the volumetric strain plotted against the average vertical strain. Generally, all the specimens show a large volumetric compression at the initial vertical strain, which then progress to a threshold compression, followed by dilation at subsequent loading cycles. With the volumetric strain, the fouled ballast with the highest *VCI* begins to dilate earlier than the specimens with a lower *VCI*. While the 70% *VCI* fouled ballast shows dilation occurring at approximately $(\varepsilon_v)_{avg} = 1.2\%$, the 20% *VCI* fouled ballast and fresh ballast starts dilating at $(\varepsilon_v)_{avg} = 1.7\%$ and $(\varepsilon_v)_{avg} = 2.2\%$, respectively. This premature dilation is one of the important signs associated with track instability (Indraratna *et al.* 2011b). Fouled specimens reinforced with geogrid show a similar trend of volumetric behaviour compared to unreinforced ballast, except that the maximum values of $(\varepsilon_v)_{avg}$ for ballast reinforced with geogrid are somewhat smaller than those for unreinforced ballast. These differences

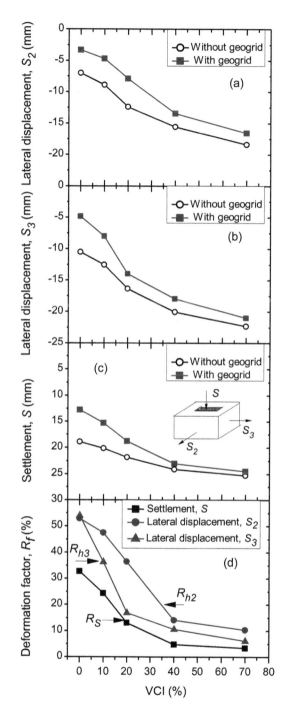

Figure 7.17 Variations of final deformation of fresh and fouled ballast with and without geogrid, with *VCI*: (a) lateral displacement S_2; (b) lateral displacement S_3; (c) settlement S; (d) relative deformation factor, R_f

(data source: Indraratna *et al.* 2013a – with permission from ASCE)

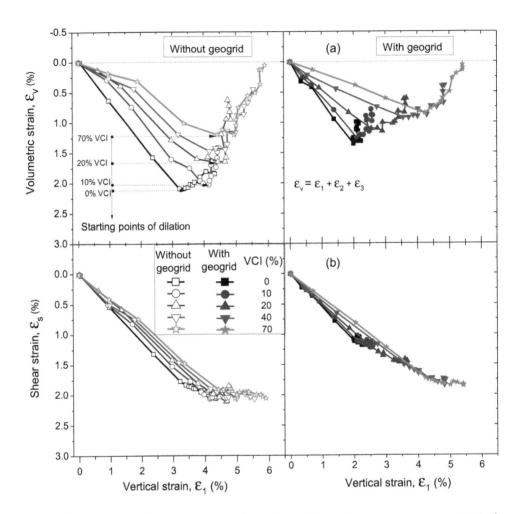

Figure 7.18 Variations of average volumetric strain and shear strain versus average vertical strain of fresh and fouled ballast with and without geogrid for various *VCI*

(after Indraratna *et al.* 2013a– with permission from ASCE)

are mainly attributed to additional interlocking provided by the geogrid which confines and restrains the ballast from moving freely.

The average shear strain response $(\varepsilon_v)_{avg}$ versus average vertical strain $(\varepsilon_1)_{avg}$ for fresh and fouled ballast is also presented in Figure 7.18b. The data show that geogrid reduces the average shear strain of fresh and fouled ballast assemblies for any given *VCI*. The shear strain for all the specimens increases significantly at the beginning, but it increases slowly with a subsequent increase in the vertical strain.

7.8.7 Maximum stresses and ballast breakage

The total stress is measured by pressure plates. The pressure plate size (230 mm in diameter by 12 mm thick) used in the laboratory is considered large enough to measure the average stress of a granular material and stiff enough to prevent any damage due to very high stress at the contact points. The pressure plates are calibrated using similar interfaces via an Instron compression machine and are found to provide the average contact stress due to granular material. Consistent results for these pressure plates were also obtained in three track sites in Singleton, Bulli and Sandgate in the state of NSW (Indraratna *et al.* 2010 and 2014).

Figure 7.19a presents a comparison of the maximum stresses at the sleeper–ballast and ballast–sub-ballast interfaces of fouled ballast with $VCI = 40\%$. As expected, the geogrid placed between the layers of ballast and sub-ballast results in a slight decrease in maximum stress compared to the unreinforced ballast assembly. It is seen that at the level of fouling $VCI = 40\%$, the reduction in stress attributed to the geogrid is not significant. This is reflected in Figure 7.19b, which shows that the reduction in breakage at 40% VCI is not very much. At the first 50,000 cycles, while the maximum vertical stress at sleeper–ballast increases, the maximum stress at the ballast–sub-ballast interface decreases. The lower stress at the sleeper–ballast interface within the initial 30,000 cycles can be attributed to inter-particle contacts that may not have been fully developed. Subsequent load cycles would increase the inter-particle contacts through densification.

Under cyclic loading, ballast deteriorates due to the breakage of sharp corners and attrition of asperities, apart from particles splitting at high contact pressure (Indraratna *et al.* 2011b; Lackenby *et al.* 2007). The amount (by mass) of broken ballast can be determined by sieving it before and after every test, and then quantifying the different particle size distribution curves. Indraratna *et al.* (2005) proposed a ballast breakage index (BBI) for quantifying ballast breakage. According to ballast samples obtained at Bulli (NSW) prior to track maintenance, the *BBI* is in the range of 8–11%, which is similar to the values obtained in the laboratory. The *BBI* of fresh and fouled ballast with and without geogrid is presented in Figure 7.19b. It can be seen here that ballast reinforced with geogrid shows a significantly reduced breakage compared to an unreinforced specimen of ballast when the VCI < 40%. This reduction of *BBI* due to including geogrid is primarily reflected by the reduced maximum stresses shown in Figure 7.19a, which would also imply reduced inter-particle contact stresses. The ability of geogrid to reduce the breakage of fresh ballast is very notable but as the *VCI* increased beyond 40%, the ability of geogrid to reduce degradation became marginal. It was also observed that the *BBI* decreased considerably with an increase of *VCI* for both reinforced and unreinforced assemblies of ballast. This was primarily attributed to the coal fines occupying the ballast voids and acting as a cushioning layer, leading to diminished inter-particle contact stresses and associated breakage.

7.8.8 Proposed deformation model of fouled ballast

Various researchers have attempted to model the settlement of fresh ballast empirically (Indraratna *et al.* 2011b; Raymond and Bathurst 1994; Shenton 1984), but their empirical equations were primarily applied to fresh ballast because they could not consider the rate at which ballast deteriorates and fouls during cyclic loading. Therefore, based on the data measured experimentally, this is the first attempt to propose an empirical equation to predict track settlement (*S*) by considering the degree of fouling (*VCI*) as defined by:

$$S = a + \frac{b}{1-VCI} log_{10} N \tag{7.12}$$

where S is the settlement, VCI is the void contaminant index ($0 \leq VCI < 1$, a and b are empirical coefficients depending on VCI and N is the number of load cycles. The parameters

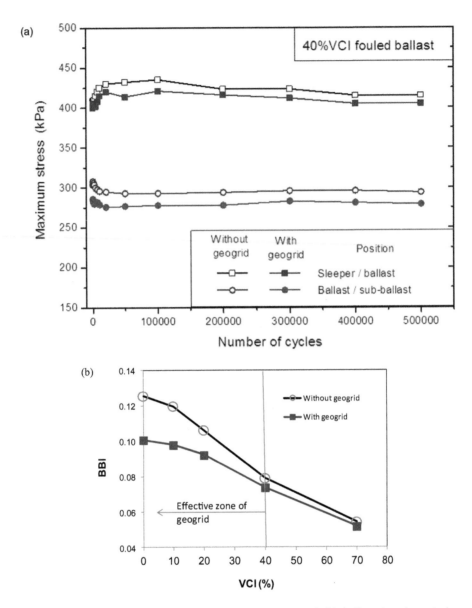

Figure 7.19 (a) Variations of maximum vertical stresses; and (b) ballast breakage index of ballast assemblies with and without geogrid

(data source: Indraratna *et al.* 2013a – with permission from ASCE)

a and *b* can vary with the subgrade characteristics, including thickness and stiffness, fouling materials and ballast gradation. This is a limitation of the proposed equation because the rigid bottom in the laboratory test setup differs from the subgrade in the field and only the commonest ballast (latite basalt) and the most common fouling material (coal) in Australian freight tracks are tested.

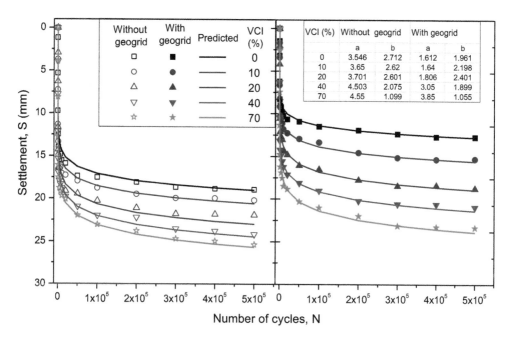

Figure 7.20 Comparisons of ballast settlement at varying *VCI* with/without geogrid inclusion measured experimentally and predicted

(data source: Ngo 2012 – PhD thesis – UOW Library)

A comparison of ballast settlement with and without geogrid at varying *VCI* is compared to the results based on Equation 7.12 and is shown in Figure 7.20. The predicted settlements agree with data measured experimentally. The empirical values of *a* and *b* at varying *VCI* were also tabulated in Figure 7.20. From a practical perspective, the proposed Equation 7.12 can help designers to predict track settlement and simultaneously consider ballast fouling. It is recommended that laboratory tests to be carried out to determine those empirical parameters if the loading and fouling materials differ from those used in this study.

Chapter 8

UOW – constitutive model for ballast

8.1 Introduction

Researchers and practitioners have long recognised that the ballast bed accumulates plastic deformation under cyclic loading. Despite this, little or no effort has been made to develop realistic constitutive stress–strain relationships, particularly modelling plastic deformation and particle degradation of ballast under cyclic loading. In case of railway ballast, the progressive change in particle geometry due to internal attrition, grinding, splitting and crushing (i.e. degradation) under cyclic traffic loads further complicates the stress–strain relationship. There is a lack of constitutive model, which includes the effect of particle breakage during shearing. In this study, a new stress–strain and particle breakage model has been developed, first for monotonic loading, and then extended for the more complex cyclic loading.

8.2 Stress and strain parameters

To develop a constitutive stress–strain and particle breakage model in a generalised stress space, a three-dimensional Cartesian coordinate system ($x_j, j = 1,2,3$) was used to define the stress and strains in ballast. Since ballast is a free draining granular medium, all the stresses used in the current model are considered to be effective.

The following general equations are being used in the model formulation:

$$q = \sigma_1 - \sigma_3 \tag{8.1}$$

$$p = \frac{1}{3}(\sigma_1 + 2\sigma_3) \tag{8.2}$$

$$\varepsilon_s = \frac{2}{3}(\varepsilon_1 - \varepsilon_3) \tag{8.3}$$

$$\varepsilon_v = \varepsilon_1 + 2\varepsilon_3 \tag{8.4}$$

The total strains ε_{ij}, are usually decomposed into elastic (recoverable) and plastic (irrecoverable) components ε^e_{ij} and ε^p_{ij}, respectively:

$$\varepsilon_{ij} = \varepsilon^e_{ij} + \varepsilon^p_{ij} \tag{8.5}$$

where the superscript e denotes the elastic component, and p represents the plastic component. Accordingly, the strain increments are also divided into elastic and plastic components:

$$d\varepsilon_{ij} = d\varepsilon^e_{ij} + d\varepsilon^p_{ij} \tag{8.6}$$

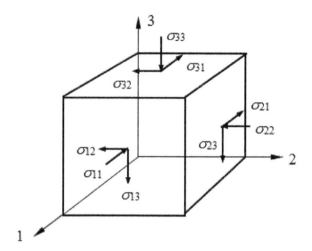

Figure 8.1 Three-dimensional stresses and index notations

Similarly, the increments of strain invariants are also separated into elastic and plastic components:

$$d\varepsilon_s = d\varepsilon_s^e + d\varepsilon_s^p \tag{8.7}$$

$$d\varepsilon_v = d\varepsilon_v^e + d\varepsilon_v^p \tag{8.8}$$

The elastic components of a strain increment can be computed using the theory of elasticity, where the elastic distortional strain increment $(d\varepsilon_s^e)$ is given by:

$$d\varepsilon_s^e = \frac{dq}{2G} \tag{8.9}$$

where G is the elastic shear modulus.

The elastic volumetric strain increment $d\varepsilon_v^e$, can be determined using the swelling/recompression constant κ, and is given by (Indraratna et al. 2011b):

$$d\varepsilon_v^e = \frac{K}{1+e_i}\left(\frac{dp}{p}\right) \tag{8.10}$$

where e_i is the initial void ratio at the start of shearing.

In this model (non-capped), the yield loci are represented by constant stress ratio (η = constant) lines in the p–q plane, inspired by Pender (1978); this is presented in Figure 8.2. The yield locus moves kinematically along with its current stress ratio as the stress changes. Mathematically, the yield function f, specifying the yield locus for the current stress ratio η_j, was expressed by Pender (1978) as:

$$f = q - \eta_j p = 0 \tag{8.11}$$

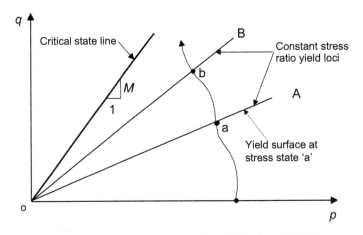

Figure 8.2 Yield loci represented by constant stress ratio lines in p, q plane

The following set of equations was used in the model calculation. More details of derivations are presented in Indraratna *et al.* (2011b).

Flow rule:

$$\frac{d\varepsilon_v^p}{d\varepsilon_s^p} = \frac{9(M-\eta)}{9+3M-2\eta^*M} + \left(\frac{B}{p}\right)\left[\frac{\chi+\mu(M-\eta^*)}{9+3M-2\eta^*M}\right] \tag{8.12}$$

where
M = critical state friction ratio, $M = 6sin\phi_f/(3- sin\phi_f)$
η = stress ratio
$\eta^* = \eta\times(p/p_{cs})$

$$B = \frac{\beta}{\ln\left(\frac{p_{cs(i)}}{p_{(i)}}\right)}\left[\frac{(9-3M)(6+4M)}{6+M}\right] = \text{constant} \tag{8.13}$$

χ and μ are two material constants relating to the rate of ballast breakage:

$$d\varepsilon_s^p = \frac{2\alpha\kappa\left(\frac{p}{p_{cs}}\right)\left(1-\frac{p_{o(i)}}{p_{cs(i)}}\right)(9+3M-2\eta^*M)\eta d\eta}{M^2(1+e_i)\left(\frac{2p_o}{p}-1\right)\left[9(M-\eta^*)+\frac{B}{p}\{x+\mu(M-\eta^*)\}\right]} \tag{8.14}$$

α is a model constant relating to the initial stiffness of ballast, and $p_{o(i)}$ and $p_{cs(i)}$ are the initial values of p_o and p_{cs}, respectively. p_{cs} is the value of p on the critical state line corresponding to the current void ratio (Fig. 8.3). Thus, $p_{cs} = exp[(\Gamma - e)/\lambda_{cs}]$, Γ = void ratio on the CSL at p = 1, and λ_{cs} is the slope of the projection of CSL on the $e - lnp$ plane, p_o is the value of p at the intersection of the undrained stress path with the initial stress ratio line (Fig. 8.4)

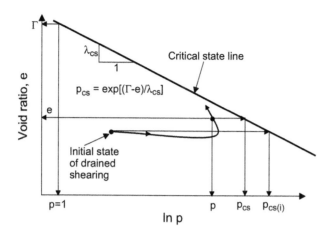

Figure 8.3 Definition of p_{cs} and typical e-ln$p$ plot in a drained shearing (adopted from Indraratna *et al.* 2011b)

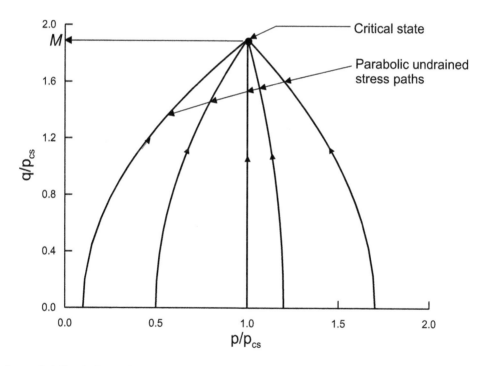

Figure 8.4 Parabolic undrained stress paths (adopted from Indraratna *et al.* 2011b)

8.2.1 Determination of model parameters

The monotonic shearing model contains 11 parameters that can be evaluated using the results of conventional drained triaxial tests and the measurements of particle breakage, as explained in this section. The critical state parameters (M, λ_{cs}, Γ and κ) can be determined from a series of drained triaxial compression tests conducted at various effective confining pressures. The slope of the line connecting the critical state points in the p–q plane gives the value of M, and that in the e-$\ln p$ plane gives λ_{cs}. The void ratio (e) of the critical state line at $p = 1$ kPa is the value of Γ. The parameter κ can be determined from an isotropic (hydrostatic) loading–unloading test with the measurements of volume change. The slope of the unloading part of isotropic test data plotted in the e-$\ln p$ plane gives the value of κ. The elastic shear modulus G can be evaluated from the unloading part of stress–strain (q-ε_s) plot in triaxial shearing.

The model parameter β (Equation 8.13) can be evaluated by measuring the particle breakage (B_g) at various strain levels, as explained in Figure 8.5.

The parameters θ and υ can be determined by re-plotting the breakage data shown in Figure 8.6 as $\ln[p_{cs(i)}/P_{(i)}]B_g$ versus ε_s^p (see Fig. 8.7), and finding the coefficients of the non-linear function that best represent the test data.

The parameters χ and μ can be evaluated by plotting the rate of particle breakage data in terms of $\ln[p_{cs(i)}/P_{(i)}]dB_g/d\varepsilon_s^p$ versus (M-η^*) (see Fig. 8.8) and determining the values of the intercept and slope of the best-fit line.

The parameter α is used in the current model to match the initial stiffness of the analytical predictions with the experimental results and can be evaluated by a regression analysis or a trial and error process that compares the model predictions with a set of experimental data.

8.2.2 Application of the UOW constitutive model to predict stress–strain responses

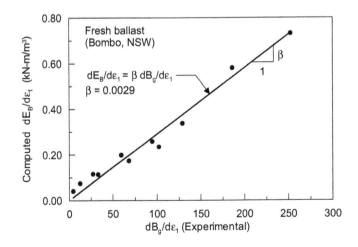

Figure 8.5 Relationship between the rate of energy consumption and rate of particle breakage

(data source from Salim 2004 – PhD thesis – UOW Library)

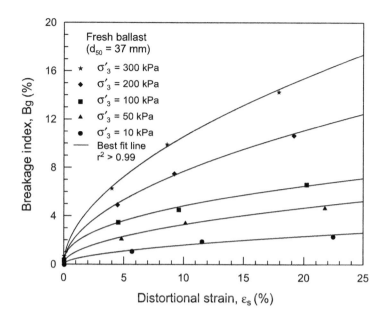

Figure 8.6 Variation of particle breakage of fresh ballast with distortional strain and confining (data source from Salim 2004 – PhD thesis – UOW Library)

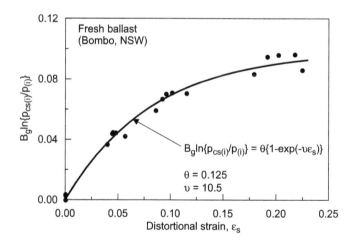

Figure 8.7 Modelling of ballast breakage during triaxial shearing (data source from Salim 2004 – PhD thesis – UOW Library)

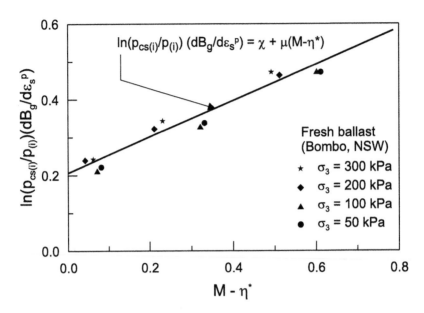

Figure 8.8 Modelling the rate of ballast breakage
(data source from Salim 2004 – PhD thesis – UOW Library)

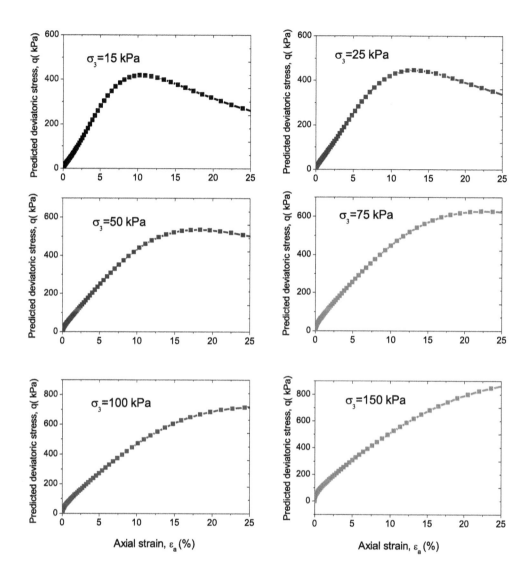

Figure 8.9 Predicted deviatoric stress versus axial strain of ballast subject to various confining pressures, ranging from 15–100 kPa

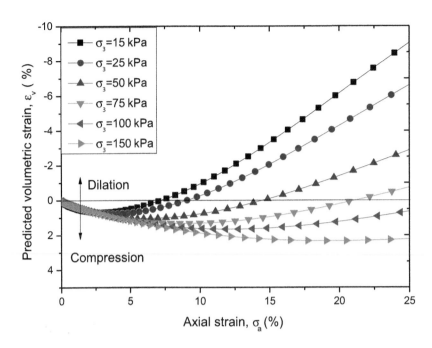

Figure 8.10 Predicted volumetric strain versus axial strain of ballast subject to various confining pressures, ranging from 15–100 kPa

Chapter 9

Sub-ballast and filtration layer – design procedure

9.1 Introduction

The hydraulic excitation in railway substructure stems from excess pore water pressure developing within the soft and saturated subgrades due to dynamic train loading. This pore water pressure generates a significant seepage force, which may transport the erodible fine particles of subgrade soil up towards the railway track. However, this problem may be avoided by placing a properly designed sub-ballast filter layer (Indraratna et al. 2012b). Two basic requirements for protective sub-ballast filters to be safe and effective include (i) retention, i.e. their ability to retain the erodible fines from the subgrade soil without clogging, and (ii) permeability, i.e. free dissipation of excess pore water pressure with adequate hydraulic conductivity ($>10^{-3}$ mm sec^{-1}). The retention requirement would also ensure that the sub-ballast filter is internally stable, so the finer fractions from the sub-ballast filter itself remains intact. The existing design criteria based on particle size distribution (PSD) of subgrade soil alone are those formulated for static conditions, so they may not always ensure a safe and effective filter for severe cyclic conditions (Indraratna et al. 2015a). Therefore, there is an urgent need for a railway-specific procedure for sub-ballast filter design. In this chapter, step-by-step design procedures are proposed for selecting and assessing the internal stability of granular sub-ballast filters in practice under cyclic loading. This chapter is divided into two main sections; namely (i) the proposed approach for practical filter design in the form of visual guidelines, and (ii) design examples to demonstrate the proposed filter design approach under cyclic loading.

The existing filter design and internal stability assessment criteria are generally similar in approach, so they possess the same limitations. For instance, the filter design approach of the International Commission on Large Dams, ICOLD (1994), recommends the use of a smaller base particle size (d_{50}) than the criteria of the Natural Resources Conservation Service, NRCS (1994), Indraratna et al. (2007), or Raut and Indraratna (2008), who recommend d_{85}, $d_{85,SA,}$ and d_{85}^*, respectively. Notably, d_{85}, $d_{85,SA}$ and d_{85}^* represent base particle sizes at 85th percentile finer by mass, by the surface area and of re-graded base soil, respectively. Nevertheless, these approaches may be conservative (safe) for some base soils and non-conservative (unsafe) for others (Raut 2006).

As discussed elsewhere by Israr et al. (2016), the existing criteria for examining the internal stability of filters generally relies on the shape and width of their PSD curve (Li and Fannin 2008; Chapuis 1992). These criteria are insensitive to the relative density (R_d) and cyclic loading on the filtration of soils, where agitation and the development of excess pore pressure under cyclic loading can induce premature seepage failures such as suffusion, piping and heave (Indraratna *et al.* 2017a). Thus, in order to minimise the risks associated with omitting these factors and to increase the longevity and durability of practical sub-ballast filters in severe dynamic conditions, enhanced filter design and stability assessment guidelines are imperative.

9.2 Requirements for effective and internally stable filters

The practical design of internally stable and effective filters involves the following steps:

Step 1:   Selection of PSD of effective protective filter to retain the given base soil:

  • Cyclic loading condition: $(D_{c35}/d_{85}) \leq 3\text{–}4$ (Trani and Indraratna 2010)

Step 2:   Geometrical assessment of internal instability potential for filters selected in step 1:

  • Cyclic loading condition: $\left(D_{c35}^{c,\,loosest} / d_{85,\,SA}^{f}\right) \leq 1$ (Israr and Indraratna 2017)

where D_{c35} and $D_{c35}^{c,\,loosest}$ define controlling constriction size at 35% finer by surface area for the filter at a given R_d and at $R_d = 0\%$, respectively, while d_{85} and $d_{85,\,SA}^{f}$ represent the base particle sizes at 85% finer by mass and by surface area, respectively. The following design charts can be used for the convenience of design practitioners.

   Figure 9.1 shows the design procedure proposed for the selection of safe and effective granular filters under cyclic loading, adopted from Israr (2016). This overall filter selection procedure involves steps which are similar to those recommended by the existing design criteria (e.g. NRCS 1994 and ICOLD 1994). These criteria would give an allowable selection band that may contain a finite number of tentative PSD curves (i.e. effective/ ineffective and internally stable/ unstable). Based on the PSD of subgrade material (base soil) to be protected, the choice of filter gradations is made from this allowable band. The selected filter gradations are then examined to determine whether or not they can protect the given base soil and remain internally stable against seepage-induced failures such as heave, piping and suffusion. It is noted in Figure 9.1 that parameters D_5, D_{15} and D_{100} are particle sizes at 5%, 15% and 100% finer for coarse fraction from PSD by mass (mm); C_u is the uniformity coefficient.

   The selected sub-ballast filter gradations should be assessed for potential internal instability by the modified combined particle and constriction size distribution (CP–CSD) based criterion of Israr and Indraratna (2017), as depicted in Figure 9.2. Note that $(H/F)_{in}$ is Kenney and Lau's (1985) stability index; R_d is relative density (%).

9.3 Filter design procedure

The current procedure for selecting a sub-ballast filter involves the following steps.

9.3.1 Internal stability of subgrade

The internal stability of the subgrade (base soil) may not be a major concern in filtration, but a prompt assessment may be made based on the secant slope of the particle size distribution curve. For instance, the minimum secant slope S of the log-linear PSD curve should be greater than 1.6 times the corresponding percentage finer by mass F (Chapuis 1992; Kenney and Lau 1985):

$$S - 1.6 \times F \geq 0 \qquad\qquad (9.1)$$

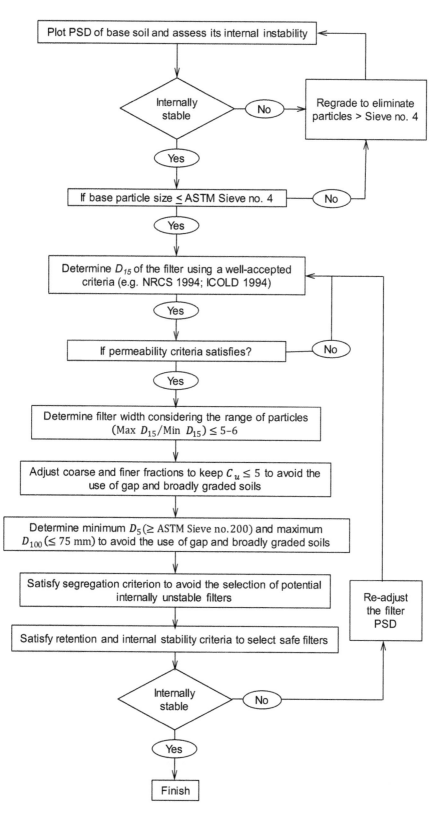

Figure 9.1 Practical implications of findings from this study: unified guidelines for selecting safe and effective granular filters under both static and cyclic conditions

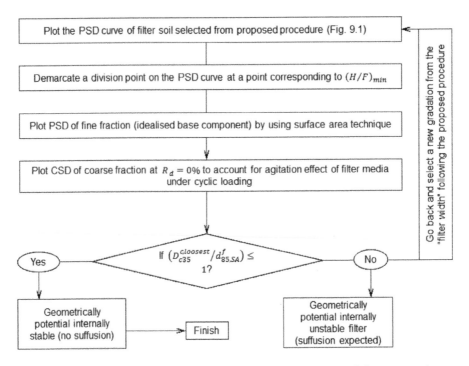

Figure 9.2 Illustration of procedure for a geometrical assessment of the potential internal instability of granular filters

Nevertheless, a more accurate assessment of potential internal instability may still be made through the constriction size distribution (CSD) based criterion (Indraratna *et al.* 2015a):

$$\left(D_{c35}^{c} / d_{85,\,SA}^{f} \leq \right) 1 \tag{9.2}$$

9.3.2 Re-grading subgrade

Locke *et al.* (2001) observed that only base soil particles smaller than 4.75 mm would be eroded due to seepage flow, and therefore a base soil with particles larger than 4.75 mm should be re-graded to consider only for the particles smaller than 4.75 mm. In essence, for soil with particle size $d_i \geq 4.75$ ($i = 1–100$), the maximum particle size $d_{100} = 4.75$ mm may be considered and the re-graded d_{85} should be used for the design allowable filter band width.

9.3.3 Selection of capping (sub-ballast) band and PSD of filter

The following tables may be used to determine allowable maximum and minimum D_{15} of the filter gradation (i.e. particle size corresponding to 15% finer by mass).

Table 9.1 Design criteria for maximum D_{15}

Base soil category	% fines (<0.075mm)	Base soil description	Maximum D_{15} (mm)
1	86–100	Fine silt and clays	$\leq 9d_{85}$
2	40–85	Sands, silts, clays and silty and clayey sand	≤ 0.7
3	15–39	Silty and clayey sands and gravel	≤ 0.7
4	0–14	Sands and gravel	$\leq 4d_{85R}$

Table 9.2 Design criteria for minimum D_{15}

Base soil category	Maximum D_{100} (mm)	Minimum D_{15} (mm)
All	≤ 75	$\geq$ ASTM #200 sieve size

To avoid segregation during placement, the following guidelines are also recommended:

Table 9.3 Segregation criteria

Base soil category	D_{10} (mm)	Minimum D_{90} (mm)
All	<0.5	20
	0.5–1.0	25
	1.0–2.0	30
	2.0–5.0	40
	5.0–10.0	50
	> 10.0	60

As general guidelines, the width of a filter band may be kept around 5 with the gradations at the boundaries having a coefficient of uniformity $C_u < 5$ (i.e. $C_u = D_{60}/D_{10}$; D_{10} and D_{60} = filter particle sizes at 10% and 60% finer by mass, respectively). Furthermore, the selected filter should offer adequate permeability ($\approx 10^{-3}$ mm sec^{-1}), which may be ensured by satisfying the Terzaghi's permeability criterion (Raut 2006):

$$\left(D_{15} / d_{15}\right) \geq 4 \tag{9.3}$$

The upper-case D denotes the size of the filter particle and the lower-case d is the size of the base particle, while the subscripts refer to the percentage of particles that are finer than that size. It is noted that the Terzaghi criteria impose two requirements, which suggests that a suitable filter must be fine enough to retain the base soil and coarse enough to drain the seepage water.

9.3.4 Internal stability of capping

Following the procedure outlined in the previous section, the potential internal instability of filter gradation may be assessed promptly using the following criterion:

$$S - 1.6 \times F \geq 0 \tag{9.4}$$

A detailed assessment of the potential internal instability of sub-ballast filter is recommended through the following CSD-based criterion (Israr and Indraratna 2017):

$$\left(D_{c35}^{c,\,loosest} \,/\, d_{85,\,SA}^{f} \leq \right)1 \tag{9.5}$$

9.3.5 Application of CSD-based retention criterion

The following CSD-based criterion of Trani and Indraratna (2010) can be used to accurately mimic the effectiveness of a sub-ballast filter:

$$\left(D_{c35} \,/\, d_{85} \right) \leq 3 - 4 \tag{9.6}$$

9.3.6 Thickness of a sub-ballast filter

To allow for variations in construction and subsequent compression under train loading, a compacted sub-ballast filter layer should have a nominal thickness of at least 150 mm (AREMA 2003; Israr 2016; Trani 2009). Given that the sub-ballast filter layer also serves as a load bearing capping layer to safely transfer the external loads from ballast to subgrade, a minimum thickness of 150 mm would be sufficient (Israr and Indraratna 2017; Israr et al. 2016; Trani and Indraratna 2010). Nonetheless, the following general criterion may be used:

$$h_{sub}\left(mm\right) = \begin{cases} 6 \times D_{100} & D_{100} \geq 25 \text{ mm} \\ 150 & D_{100} < 25 \text{ mm} \end{cases} \tag{9.7}$$

where h_{sub} = thickness of sub-ballast filter layer.

Chapter 10

Practical design examples

Several practical track design examples are provided here; each example focuses on a particular design feature. These specific problems provide information on the input and output of various case studies as a valuable guide for real world applications. Designers are encouraged to study these examples and with further modifications may use them to solve similar problems, while realising that these examples do not explore all the full functions and design encountered in the practice. The main features of each worked-out example are summarised as follows:

- Example 1: Calculate the bearing capacity of ballasted tracks (underneath the sleeper)
- Example 2: Determine the thickness of granular layer
- Example 3: Ballast fouling and implications on drainage capacity, train speed
- Example 4: Use of geosynthetics in ballasted tracks
- Example 5: Evaluation of track modulus and settlement
- Example 6: Determine the friction angle of fouled ballast
- Example 7: Determine the settlement of fouled ballast
- Example 8: Calculate the ballast breakage index (BBI)
- Example 9: Effect of the depth of subgrade on determining thickness of granular layer
- Example 10: Design sub-ballast/capping as a filtration layer for track

10.1 Worked-out example 1: calculate the bearing capacity of ballasted tracks

10.1.1 Design parameters

- Rail and sleeper properties: sleeper width, B = 0.25(m); sleeper length, L = 2.4 (m); spacing of sleeper, s = 0.61(m).
- Ballast properties: unit weight of ballast, γ = 15.6 (kN m^{-3}); friction angle of ballast, ϕ = 48°
- Traffic condition: static wheel load, P_s = 12.5 (ton) – 25-ton axle load; wheel diameter, D = 0.97 (m); train velocity, V = 85 (km h^{-1}), factor of safety, FOS = 1.2.

10.1.2 Calculation procedure

Step 1: Calculate the passive Rankin earth pressure coefficient K_p (Assuming that passive conditions are fully mobilised):

$$K_p = \frac{1 + \sin(48)}{1 - \sin(48)} = 6.786$$

Step 2: Calculate the bearing capacity factors N_q and N_γ:

$$N_q = K_p \times e^{(\pi \times \tan \phi)} = 222.3$$

$$N_\gamma = \left(N_q - 1\right) \tan\left(1.4\phi\right) = (2223 - 1) \times \tan(1.4 \times 48) = 526.45$$

Step 3: Calculate the shape factor S_γ:

$$S_\gamma = 1 + 0.1 K_p \left(B/L\right) = 1 + \left(0.1 \times 6.786 \times \left[\frac{0.25}{2.4}\right]\right) = 1.071$$

Step 4: Ultimate and allowable bearing capacity calculation:
Ultimate bearing capacity q_{ult}:

$$q_{ult} = N_\gamma S_\gamma (0.5 \gamma B - \Delta u) = 526.45 \times 1.071 \times (0.5 \times 15.6 \times 0.25) = 916 \text{kPa}$$

$$\text{Allowable bearing capacity} = \frac{q_{ult}}{FOS} = \left(\frac{905.5}{1.2}\right) = 754.6 \text{kpa}$$

Step 5: Dynamic amplification factor (*IF*) calculation (AREA (1974) method is used):

$$IF = 1 + 5.21 \frac{V}{D} = 1 + 5.21 \ x \ 85/970 = 1.457$$

Step 6: Rail seat load method calculation (AREA (1974) method is used):

$$q_r = D_f P_s$$

where P_s is the input axle load:

$$D_f = 0.45 + 5.77 \times 10^{-4} S_s = 0.45 + (5.77 \times 10^{-4} \times 610) = 0.802$$

$$\text{Maximum static stress} = \frac{\text{Retail seat load}}{\text{Effective sleeper area}}$$

Effective sleeper area = Sleeper width $\times$ sleeper length/3

Therefore,

$$\text{Maximum Static Stress} = \frac{0.802 \times 122.6}{0.25 \times (2.4/3)}$$
$$= 491.7 \text{kPa}$$

$$\text{Equivalent Dynamic Stress} = IF \times \text{Maximum static Stress}$$
$$= 491.63 \times 1.457$$
$$= 716.2 \text{kPa}$$

Compare this maximum dynamic stress with the allowable bearing capacity. As long as the equivalent dynamic stress is less than the allowable bearing stress, the load-bearing capacity criteria is satisfied. In this example, the bearing capacity of ballast is satisfied.

10.2 Worked-out example 2: determine the thickness of granular layer

This example is similar to the one presented by Li and Selig (1998a, b). The track is built on a homogeneous and uniform clay subgrade. The track is subjected to regular heavy wheel loads, the details of which are summarised in Table 10.1.

10.2.1 Calculation procedure

Step 1: Dynamic amplification factor (IF) calculation (AREA (1974) method is used):

$$IF = 1 + 5.21 \frac{V}{D} = 1 + 5.21 \frac{64}{970} = 1.34$$

Step 2: Dynamic wheel load (P_d) calculation:

$$P_d = 1.34 \times 173 = 232.47 \text{ kN}$$

Step 3: Number of load cycle (N) calculation:

$$N = \frac{(60 \times 10^6) \times 9.81}{8 \times 173} = 425,289$$

Table 10.1 Input parameters for determining the thickness of the granular layer based on the performance of the subgrade (Li and Selig 1998b)

Design parameters	Values
Design criteria	Allowable subgrade plastic strain for the design period, ε_{pa} = 2%
	Allowable settlement of subgrade in design period, ρ_a = 25 mm
	Minimum granular layer height = 0.45 m
	Impact factor method = AREA (1974)
	Subgrade capacity method = Li and Selig (1998a, b)
Rail and sleeper properties	Not needed
Traffic conditions	Static wheel load P_s = 173 kN
	Velocity V = 64 km h
	Design tonnage = 64 MGT
	Wheel diameter, D = 0.97 m
Granular material characteristics	Resilient modulus, $E_b = E_c = E_s$ = 276 MPa
Subgrade soil characteristics	Soil type: CH (fat clay)
	Soil compressive strength, σ_s = 90 kPa
	Subgrade modulus, E_s = 14 MPa
	Thickness = 1.5 m

Step 4: Select the values of a, m and b for CH soil from Table 4.1:

$a = 1.2$; $m = 2.4$; $b = 0.18$

Step 5: Calculation for the first design procedure (preventing local shear failure of subgrade).

Step 5.1: Calculate allowable deviator stress on subgrade (σ_{da}) using Equation 4.4:
$\varepsilon_p = 2\%$ and $\sigma_s = 90$ kPa is used from the input parameters

$$\sigma_{da} = \sigma_s\left[\left(\frac{\varepsilon_p}{aN^b}\right)^{(1/m)}\right] = 90\left[\left(\frac{2}{1.2\times425289^{0.18}}\right)^{(1/2.4)}\right] = 42.1\text{kPa}$$

Step 5.2: Calculate the strain influence factor (Iε) from Equation 4.5:

$$I\mu = \frac{42.1\times0.645}{232.47} = 0.117$$

Step 5.3: Determine the (H/L) from Figure 4.3 corresponding to the granular material modulus $E_s = 276$ MPa and subgrade modulus 14 MPa;

H/L = 5.07

H = $5.07\times0.152 = 0.771$ m

Step 6: Calculation for the second design procedure (preventing excessive plastic deformation of the subgrade layer).

Step 6.1: Calculate the deformation influence factor I_{pa}:

$$I_{pa} = \frac{\dfrac{(25\times10^{-3})}{0.152}}{1.2\times\left(\dfrac{232.47}{90\times0.645}\right)^{2.4}\times425289^{0.18}}\times100 = 0.048$$

Step 6.2: Determine the (H/L) from Figure 4.4:

H/L = 4.74

H = $4.74\times0.152 = 0.721$ m

Step 7: Design thickness of the granular layer is selected as the higher values obtained from Step 5.3 and Step 6.2, that is, H_{design} = 771 (mm).

10.3 Worked-out example 3: ballast fouling and implications on drainage capacity, train speed

This section provides a practical example of the quantification of ballast fouling and the implications of fouling on the drainage capacity of ballast and potentially reduced train speed due to fouling.

10.3.1 Calculate levels of ballast fouling

Ballast fouling can be quantified using the void contaminant index (*VCI*) introduced by Tenakoon *et al.* (2012), as described in the Chapter 6, Equation 6.3:

$$VCI = \frac{(1+e_f)}{e_b} \times \frac{G_{sb}}{G_{sf}} \times \frac{M_f}{M_b} \times 100$$

Input design parameters

Fouling materials: void ratio, $e_f = 0.76$; mass, $M_f = 5$ (kg); specific gravity, $G_{sf} = 1.32$

Ballast: void ratio, $e_b = 0.75$; mass $M_b = 100$ (kg); specific gravity, $G_{sb} = 2.75$:

$$VCI = \frac{(1+0.76)}{0.75} \times \frac{2.75}{1.32} \times \frac{5}{100} \times 100 = 24.4\%$$

10.3.2 Effect of ballast fouling on track drainage

Input design parameters

- Hydraulic conductivity of fresh (clean) ballast, $k_{ba} = 406.29 \times 10^{-3}$ (m s^{-1})
- Hydraulic conductivity of fouling materials, $k_{co} = 9 \times 10^{-5}$ (m s^{-1})
- Void contaminant index, $VCI = 24.4\%$

Calculation: relative hydraulic conductivity: $\left(\dfrac{k_b}{k}\right)$

$$\left(\frac{k_b}{k}\right) = 1 + \frac{VCI}{100}\left(\frac{k_b}{k_f} - 1\right) = 1 + \frac{24.44}{100}\left(\frac{0.40629}{0.00009} - 1\right) = 1104.09$$

Check with the drainage criteria in the Table 6.3 and it falls into acceptable drainage.

10.3.3 Fouling versus train speed

Input design parameters

- Rail and sleeper properties: sleeper width, B = 0.25(m); sleeper length, L = 2.4 (m); spacing of sleeper, s = 0.61(m).
- Ballast properties: unit weight of ballast, $\gamma = 15.6$ (kN m^{-3}); friction angle of ballast, $\phi = 48°$; design ballast thickness, h = 0.3 (m); peak strength (at $\sigma_3 = 10$ *kpa*) = 280 kPa; peak strength (at $\sigma_3 = 30$ *kpa*) = 340 kPa; the values of β are given in Table 6.4.
- Traffic condition: static wheel load, Ps = 122.3 (kN); wheel diameter, D = 0.97 (m); train velocity, V = 85 (km h^{-1}), factor of safety, FOS = 1.2. Train speed for clean ballast (at $\sigma_3 = 10$ *kpa*) = 100 km h^{-1}; train speed for clean ballast (at $\sigma_3 = 30$ *kpa*) = 125 km h^{-1}.

Calculation

- Dynamic amplification factor (AREA method):

$$\sigma_3 = 10 \, kPa: \ IF = 1.536; \quad \sigma_3 = 30 \, kPa: \ IF = 1.671$$

- Dynamic stress under different σ_3 (i.e. changes in train speed):

$$\sigma_3 = 10 \, kPa: \ P_d = 248 \text{ kPa}; \quad \sigma_3 = 30 \, kPa: \ P_d = 269.7 \text{ kPa};$$

- Calculate velocity reduction factor, VRR, using Equation 6.13:

$$\text{VRR at } \sigma_3 = 10\text{kPa}$$

$$\text{VRR} = \frac{\left[\dfrac{191.16 \times (0.25 + 0.3/2)(2.4/3 + 0.3/2)}{0.5 \times 122.63}\right] - 1}{\left[\dfrac{280 \times (0.25 + 0.3/2)(2.4/3 + 0.3/2)}{0.5 \times 122.63}\right] - 1} = 0.251$$

$$\text{VRR at } \sigma_3 = 30\text{kPa}$$

$$\text{VRR} = \frac{\left[\dfrac{275.9 \times (0.25 + 0.3/2)(2.4/3 + 0.3/2)}{0.5 \times 122.63}\right] - 1}{\left[\dfrac{340 \times (0.25 + 0.3/2)(2.4/3 + 0.3/2)}{0.5 \times 122.63}\right] - 1} = 0.641$$

Maximum permissible speeds (only considering shear strength aspects) are calculated from Equation 6.11 as:

$$\text{Max permissible speed at } \sigma_3' = 10\text{kPa} = \left[\frac{q_{peak,b} \times (B + h/2)(l/3 + h/2)}{0.5 \times P_s} - 1\right]\frac{D_w}{0.0052}$$

$$= \left[\frac{280 \times (0.25 + 0.3/2)(2.4/3 + 0.3/2)}{0.5 \times 122.63} - 1\right]\frac{0.97}{0.0052}$$

$$= 137.16\text{km/h}$$

$$\text{Max permissible speed at } \sigma_3' = 30\text{kPa} = \left[\frac{340 \times (0.25 + 0.3/2)(2.4/3 + 0.3/2)}{0.5 \times 122.63} - 1\right]\frac{0.97}{0.0052}$$

$$= 206.53\text{km/h}$$

Note: "The above speeds are to be reduced further using a safety factor of 1.3–1.5 considering track degradation over time."

10.4 Worked-out example 4: use of geosynthetics in ballasted tracks

10.4.1 Design input parameters

Mean particle size of ballast, $D_{50} = 35$ (mm)
Selected aperture size (i.e. opening area) of geogrid: 40 mm × 40 mm

- Recommended aperture size of geogrid based on UOW test data:

$$A_{optimun} = 1.2 \times D_{50} = 1.2 \times 35 = 42 \ (\text{mm})$$

- Minimum aperture size of geogrid can be used (i.e. to prevent creating a slipping plane):

$$A_{minimum} = 0.95 \times D_{50} = 0.95 \times 35 = 33.25 \ (\text{mm})$$

- Maximum aperture size of geogrid can be used (i.e. to enable the interlocking with ballast aggregates):

$$A_{minimum} = 0.95 \times D_{50} = 0.95 \times 35 = 33.25 \ (\text{mm})$$

- Interface efficiency factor, corresponding to $A/D_{50} = 40/35 = 1.14$

$$\alpha = \frac{\tan(\delta)}{\tan(\phi)} = 1.12 \ > \ 1$$

Therefore, the geogrid selected is acceptable.

10.4.2 Predicted settlement of fresh ballast after N = 500,000 load cycles

Unreinforced fresh ballast (using N=500,000):

$$Settlement \ (mm) = a + b \times \ln(N) = 2.94 + 0.56 \times \ln(500,000) = 10.29 \ (mm)$$

Geogrid-reinforced fresh ballast:

$$Settlement \ (mm) = a + b \times \ln(N) = 2.09 + 0.48 \times \ln(500,000) = 8.39 \ (mm)$$

Settlement reduction factor:

$$SRF = 1 - \frac{Settlement_{reinforced}}{Settlement_{unreinforced}} = 1 - \frac{8.39}{10.29} = 0.18$$

10.4.3 Recycled ballast

Unreinforced recycled ballast:

$$Settlement \ (mm) = c + d \times \ln(N) = 9.12 + 0.53 \times \ln(500,000) = 16.07 \ (mm)$$

Geogrid-reinforced recycled ballast:

$$Settlement \ (mm) = c + d \times \ln(N) = 7.71 + 0.50 \times \ln(500,000) = 14.27 \ (mm)$$

Settlement reduction factor:

$$SRF = 1 - \frac{Settlement_{reinforced}}{Settlement_{unreinforced}} = 1 - \frac{14.27}{16.07} = 0.112$$

10.5 Worked-out example 5: evaluation of track modulus and settlement

10.5.1 Determine the overall track modulus for a given track structure with the following information

Ballast layer: thickness, H_b = 300 mm; modulus of ballast: E_b = 200 MPa
Capping layer: thickness, H_c = 150 mm; modulus of capping: E_c = 115 MPa
Structural fill: thickness, H_f = 600 mm; modulus of structure fill: E_f = 45 MPa

10.5.2 Calculation procedure

Equivalent modulus of granular layer ($\bar{E}$) as shown in Figure 10.1 is calculated using:

$$\bar{E} = \frac{H_b + H_c + H_f}{\left(\dfrac{H_b}{E_b} + \dfrac{H_c}{E_c} + \dfrac{H_f}{E_f}\right)}$$

where E_b, E_c and E_f are the values of elastic modulus of the ballast, capping and structural fill. H_b, H_c and H_f are the thicknesses of ballast, capping and structural fill, respectively:

$$\bar{E} = \frac{0.3 + 0.15 + 0.6}{\left(\dfrac{0.3}{200} + \dfrac{0.15}{115} + \dfrac{0.6}{45}\right)} = 65.06 \text{MPa}$$

Assuming the equivalent dynamic stress at the sleeper–ballast interface as 716.2 kPa (see calculations in design example 1), the average strain (ε_{ave}) of the equivalent granular medium is then calculated at the stress at sleeper–ballast interface divided by the equivalent modulus ($\bar{E}$), thus:

$$\varepsilon_{ave} = \frac{\sigma_{dyn}}{\bar{E}} = \frac{716.2}{65.06 \times 10^3} \times 100\% = 1.1\%$$

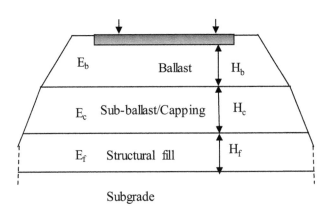

Figure 10.1 Schematic diagram of a typical track substructure

The overall elastic settlement of granular medium is then predicted using the total height (H = 1.05m):

$$S_{total} = \varepsilon_{ave} \times H = 1.1/100 \times 1.05 = 0.0116 \; (m) = 11.6 \; (mm)$$

10.6 Worked-out example 6: determine the friction angle of fouled ballast

This worked-out example presents a procedure to predict a decreased friction angle of coal-fouled ballast, and the correspondingly increased track settlement for a given level of ballast fouling based on the void contamination index (VCI).

- The degree of ballast fouling is considered for: VCI = 20%, 40%, 70% and 95%.
- The values of the apparent friction angle of fresh ballast (φ_0) are assumed to vary between 55° and 65° depending on the applied stress levels (Fig. 7.9b).

10.6.1 Calculation procedure

According to Chapter 7, the shear strength and friction angle of fouled ballast decrease steadily with an increase in the content of fines. The variations in the decrease of normalised peak shear stress with respect to changes in the VCI can then be presented by the following hyperbolic equation (Indraratna et al. 2011b):

$$\frac{\Delta \tau_p}{\sigma_n} = \frac{VCI/100}{a \times VCI/100 + b}$$

where $\Delta \tau_p$ = the shear strength reduction of ballast due to the presence of fines, σ_n = normal stress, VCI = the void contamination index, and a and b = hyperbolic constants, presented in Table 10.2.

Incorporating the Mohr–Coulomb envelope for a cohesionless material, the friction angle of fouled ballast can be approximated by:

$$\tan \phi_f = \tan \phi_0 - \frac{VCI/100}{a \times VCI/100 + b}$$

where φ_f = the peak angle of shearing resistance of fouled ballast and φ_0 = the peak angle of shearing resistance of fresh ballast. The estimated friction angle of coal-fouled ballast can then be determined, as presented in Table 10.3, Figure 10.2 and Figure 10.3.

Table 10.2 Hyperbolic empirical constants a and b obtained via laboratory tests

Normal stress (kPa)	Without geogrid		With geogrid	
σ_n	a	b	a	b
15	0.81	0.48	0.77	0.78
27	0.88	0.65	0.81	0.92
51	1.03	0.74	0.91	1.10
75	1.38	0.85	1.17	1.15

Table 10.3 Predicted friction angle of fouled ballast with and without the inclusion of geogrid

VCI (%)	Without geogrid				With geogrid (40 mm × 40mm)			
σ_n	15 kPa	27 kPa	51 kPa	75 kPa	15 kPa	27 kPa	51 kPa	75 kPa
0	65.05	62.6	61.23	56.27	69.33	65.44	63.49	60.31
20	61.29	59.36	58.41	52.86	67.66	63.37	61.49	58.11
40	58.95	57.05	55.43	50.6	66.54	62.07	60.3	56.53
70	55.32	53.35	52.8	47.73	64.75	59.46	57.71	54.23
95	53.78	52.57	52.18	46.78	63.97	58.47	56.8	53.19

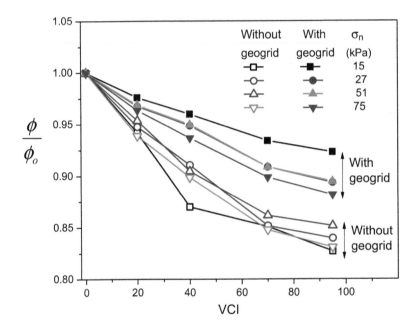

Figure 10.2 Normalised angle of shearing resistance by initial angle of shearing resistance versus VCI for ballast with and without geogrid reinforcement

10.7 Worked-out example 7: determine the settlement of fouled ballast

This worked-out example presents a procedure to predict settlement of coal-fouled ballast for a given level of fouling.

*   The degree of ballast fouling for: VCI = 20, 40, 70 and 95%.
*   The values of the apparent friction angle of fresh ballast (φ_0) are assumed to vary between 55° and 65° depending on the applied stress levels.

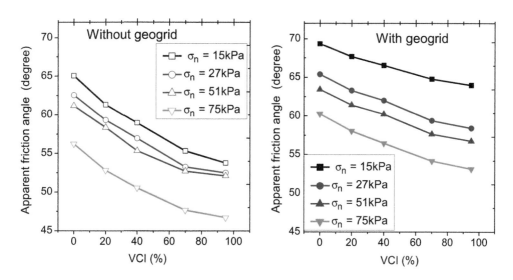

Figure 10.3 Predicted friction angle of fouled ballast with and without the inclusion of geogrid

Note: User is required to enter the degree of fouling in terms of the parameter *VCI* and the applied normal stress on ballast, σ_n. UOW test data currently incorporate the data corresponding to the applied normal stresses of 15, 27, 51 and 75 kPa based on a study carried out by Indraratna *et al.* (2011a) using a biaxial geogrid (aperture: 40 mm × 40 mm). Ideally, the designers are advised to carry out such tests to obtain the relevant parameters that are suitable for a particular track design, as the type of geogrid, ballast gradation and applied normal stresses can influence the test results.

10.7.1 Calculation procedure

Based on extensive test data measured using the large-scale track process simulation apparatus (PSTA) described in Chapter 7, an empirical equation to predict track settlement (*S*) considering the extent of fouling (*VCI*) can be defined by (Indraratna et al. 2013a):

$$S = a + \frac{b}{1-VCI} log_{10} N$$

where *S* is the settlement, *VCI* is the void contaminant index ($0 \leq VCI < 1$), *a* and *b* are empirical coefficients depending on *VCI*, and *N* is the number of load cycles. For given test conditions and materials examined in this study, the values of parameters *a* and *b* are presented in Table 10.4.

Settlement of coal-fouled ballast for a given VCI and number of load cycles (N) can then be estimated using the above equation, and their values are presented in Table 10.5 and Figure 10.4.

10.8 Worked-out example 8: calculate the ballast breakage index (BBI)

This worked-out example presents a procedure to predict ballast breakage index (BBI) of ballast for a given level of confining pressure.

Table 10.4 Empirical parameters *a* and *b* obtained from laboratory tests

VCI (%)	Without geogrid		With geogrid	
	a	*b*	*a*	*b*
0	3.546	2.712	1.612	1.961
10	3.65	2.62	1.64	2.198
20	3.701	2.601	1.806	2.401
40	4.503	2.075	3.05	1.899
70	4.55	1.099	3.85	1.055

Table 10.5 Predicted settlements of fouled ballast with and without geogrid inclusion

Load cycle	Without geogrid					With geogrid				
N	VCI = 0	VCI = 10	VCI = 20	VCI = 40	VCI = 70	VCI = 0	VCI = 10	VCI = 20	VCI = 40	VCI = 70
1	3.55	3.96	4.09	4.71	4.77	1.61	1.64	1.81	3.32	3.81
10	6.26	6.87	7.40	8.17	8.44	3.57	4.08	4.81	6.49	7.32
100	8.97	9.78	10.71	11.63	12.10	5.53	6.52	7.81	9.65	10.84
1000	11.68	12.69	14.02	15.08	15.76	7.49	8.97	10.81	12.82	14.36
2000	12.50	13.56	15.02	16.13	16.86	8.09	9.70	11.71	13.77	15.42
4000	13.31	14.44	16.01	17.17	17.97	8.68	10.44	12.61	14.72	16.48
7000	13.97	15.15	16.82	18.01	18.86	9.15	11.03	13.34	15.49	17.33
10,000	14.39	15.60	17.33	18.54	19.42	9.46	11.41	13.81	15.98	17.88
20,000	15.21	16.48	18.33	19.58	20.53	10.05	12.14	14.71	16.93	18.94
50,000	16.29	17.63	19.64	20.96	21.99	10.83	13.12	15.91	18.19	20.34
100,000	17.11	18.51	20.64	22.00	23.09	11.42	13.85	16.81	19.15	21.40
200,000	17.92	19.39	21.64	23.04	24.19	12.01	14.59	17.71	20.10	22.45
300,000	18.40	19.90	22.22	23.65	24.84	12.35	15.02	18.24	20.66	23.07
400,000	18.74	20.26	22.63	24.08	25.29	12.60	15.32	18.62	21.05	23.51
500,000	19.00	20.54	22.95	24.42	25.65	12.79	15.56	18.91	21.36	23.85

10.8.1 Calculation procedure

Indraratna *et al.* (2011b) demonstrated with experimental evidence that particle breakage increases with increasing axial strain, but at a decreasing rate, finally approaching a relatively constant value. The value of breakage index also becomes greater as the confining pressure increases. From experimental findings carried out by Indraratna *et al.* (2015b), a unified function is proposed as following to represent the particle breakage during shearing:

$$BBI = \frac{\theta_b[1-\exp(-v_b\varepsilon_s^p)]}{\omega_b \ln p_t'} \qquad (10.1)$$

where θ_b, v_b and ω_b are material constants characterising the breakage of aggregates, and p_i' is the initial effective mean stress which equals the initial track confinement. For latite basalt (Fig. 10.5), a commonly used ballast aggregate in the state of New South Wales, the values of θ_b, v_b and ω_b are recommended as: 0.33, 11.5 and 6.4. The predicted values of BBI at various track confinement are plotted in Figure 10.6 and given as data in Table 10.6.

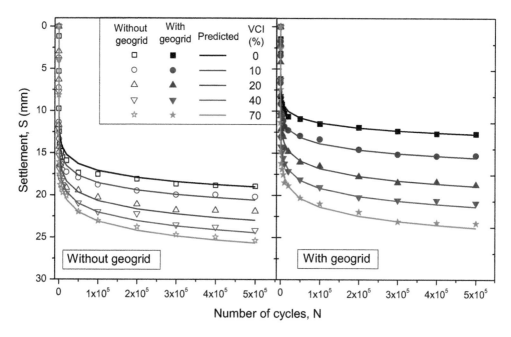

Figure 10.4 Predicted settlements of fouled ballast with number of load cycles

Note: Empirical parameters *a* and *b* can vary considerably depending on the mechanical properties of the track substructure, including thickness and modulus of each layer, nature of fouling materials, ballast gradation and parent rock type and the cyclic loading conditions. Predicted values of track settlements as given in Table 10.5 can be used in preliminary track design, incorporating the effect of ballast fouling. For more comprehensive track design, it is strongly recommended that designers should carry out independent tests to obtain the parameters that are specific for a given track condition; the proposed values by the authors only serve as preliminary guidance.

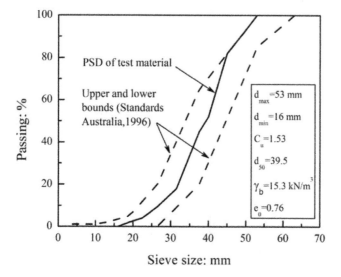

Figure 10.5 Particle size distribution of tested ballast

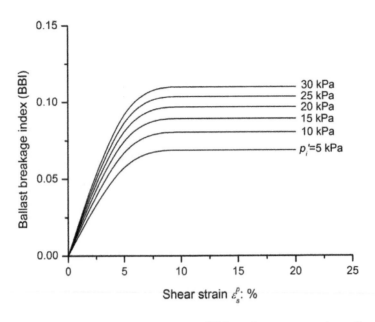

Figure 10.6 Prediction of ballast breakage index (BBI) with various track confinement

Table 10.6 Predicted values of BBI under different track confinements and shear strain

ε_s (%)	p_i: confinement (kPa)	BBI
0	5	0.0000
5	5	0.0689
10	5	0.0689
15	5	0.0689
20	5	0.0689
0	10	0.0000
5	10	0.0805
10	10	0.0805
15	10	0.0805
20	10	0.0805
0	15	0.0000
5	15	0.0894
10	15	0.0894
15	15	0.0894
20	15	0.0894
0	20	0.0000
5	20	0.0969
10	20	0.0969
15	20	0.0969
20	20	0.0969
0	25	0.0000
5	25	0.1037
10	25	0.1037
15	25	0.1037
20	25	0.1037
0	30	0.0000
5	30	0.1100
10	30	0.1100
15	30	0.1100
20	30	0.1100

10.9 Worked-out example 9: effect of the depth of subgrade on determine thickness of granular layer

This worked-out example will look at the effect of the subgrade depth in designing ballasted track substructure.

It is noted that the use of design charts introduced by Li and Selig (1998a) in determining the thickness of granular layer can sometimes be incomplete or unreliable without knowing the actual depth factor of the subgrade soil. In particular, for soft plastic clays (CH) it can make a significant difference to track response if the CH thickness at a given location is 30 m or 3 m. In addition, the use of non-destructive modulus or elastic (small strain) stiffness for a very thick soft clay layer can also be misleading.

It is therefore suggested to consider the elastic stress distribution with depth (e.g. Boussinesq's method) and calculate the effective depth of subgrade that primarily influences design, and then establish an equivalent stiffness for that depth. In this example, assuming the equivalent stress at the sleeper–ballast interface as 491.7 kPa, the stress distribution with depth is presented in Figure 10.7.

It is seen that the vertical stress decreases significantly with depth, and the effective depth of subgrade that influences the design in this given example can be taken as 4 m. Below this depth, the property of subgrade does little to affect the stress distribution and associated behaviour. Therefore, it is suggested to consider a depth of 4 m when calculating an equivalent stiffness for the subgrade using the concept of equivalent modulus, given by:

$$\bar{E} = \frac{H_b + H_c + H_s}{\dfrac{H_b}{E_b} + \dfrac{H_c}{E_c} + \dfrac{H_s}{E_s}}$$

where E_b, E_c and E_s are the elastic moduli of the ballast, capping and structural fill. H_b, H_c and H_s are the thicknesses of ballast, capping and subgrade, respectively (Fig. 10.8).

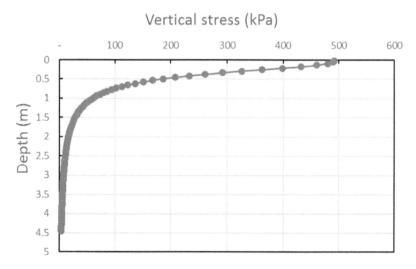

Figure 10.7 Stress distribution with depth using Boussinesq's method

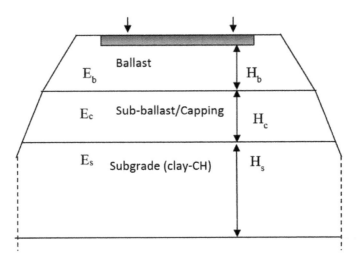

Figure 10.8 Schematic diagram of granular layers in ballasted track substructure

10.9.1 Example: Design a ballasted track substructure for a train crossing two different sections over the same highly plastic clay subgrade (CH); one section has 10 times the thickness of the other

Track section 1	Track section 2
Ballast layer: H_b = 300 mm; E_b = 200 MPa	Ballast layer: H_b = 300 mm; E_b = 200 MPa
Capping layer: H_c = 150 mm; E_c = 95 MPa	Capping layer: H_c = 150 mm; E_c = 95 MPa
Subgrade: H_s = 2 m; E_s = 20 MPa	Clay subgrade: Hs = 20 m; E_s = 20 kPa
(CH – clays with high plasticity)	(CH – clays with high plasticity)
Equivalent dynamic stress at the sleeper–ballast, σ_{dyn} = 491.7 kPa	Equivalent dynamic stress at the sleeper–ballast, σ_{dyn} = 491.7 kPa

Equivalent modulus of the subgrade layer ($\overline{E}$) for each track section is now calculated by:

- Track section 1:

$$\overline{E}_1 = \frac{H_b + H_c + H_s}{\dfrac{H_b}{E_b} + \dfrac{H_c}{E_c} + \dfrac{H_s}{E_s}} = \frac{0.3 + 0.15 + 2}{\dfrac{0.3}{200} + \dfrac{0.15}{95} + \dfrac{2}{20}} = 23.8 MPa$$

- Track section 2:

$$\overline{E}_2 = \frac{H_b + H_c + H_s}{\dfrac{H_b}{E_b} + \dfrac{H_c}{E_c} + \dfrac{H_s}{E_s}} = \frac{0.3 + 0.15 + 4}{\dfrac{0.3}{200} + \dfrac{0.15}{95} + \dfrac{4}{20}} = 21.9 MPa$$

The average strain (ε_{ave}) of the equivalent granular medium is determined as:

- Track section 1:

$$\varepsilon_{ave\_1} = \frac{\sigma_{dyn}}{\overline{E_1}} = \frac{491.7}{23.8 \times 10^3} \times 100\% = 2.06\%$$

- Track section 2:

$$\varepsilon_{ave\_2} = \frac{\sigma_{dyn}}{\overline{E_2}} = \frac{491.7}{21.9 \times 10^3} \times 100\% = 2.25\%$$

The settlement of the entire track substructure layer can be estimated as follows:

- Track section 1:

$$S_{total\_1} = \varepsilon_{ave\_1} \times H_1 = 2.06/100 \times (0.3 + 0.15 + 2) = 0.05 \ (\text{m})$$

- Track section 2:

$$S_{total\_2} = \varepsilon_{ave\_2} \times H_2 = 2.25/100 \times (0.3 + 0.15 + 4) = 0.10 \ (\text{m})$$

10.9.2 Design summary

Track section	Subgrade thickness (m)	Equivalent modulus (MPa)	Average strain (εave)	Settlement (m)
1	2	23.8	3.01	0.05
2	20	21.9	3.27	0.1

10.10 Worked-out example 10: design of sub-ballast/ capping as a filtration layer for track

Two simple design examples are presented in this section to demonstrate the proposed design procedure for sub-ballast filter

10.10.1 Design example 10.1: selection of effective granular filters effective to retain a base soil under given hydraulic conditions

Design a safe and effective railway sub-ballast filter to protect a highly dispersive silty sub-grade soil, shown in Figure 10.9, from erosion by a freight train moving up to a maximum speed of 110 km h^{-1} (i.e. 20 Hz) on a standard gauge rail-track in NSW; also determine the potential internal instability for the filter gradations selected, and assume suitable values for the missing data.

10.10.2 Sub-ballast filter design

The procedure described in Figure 9.1 is adopted here to select a filter band for cyclic conditions that requires a relaxed CSD based criterion ($D_{C35}/d_{85} \leq 3$–4). For the given R_d, the

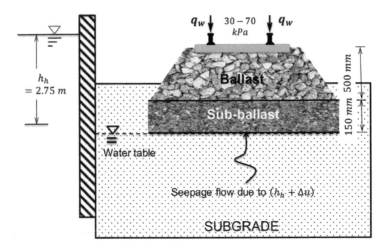

Figure 10.9 Illustration of scenarios in design examples (not scaled)

Table 10.7 Summary of calculations for geometrical design of sub-ballast filter

Filter	R_d (%)	C_u	D_{15} (mm)	D_{c35} (mm)	D_{c35}/d_{85}	Effective (Yes/No)
F1	> 95	4.7	0.3	0.057	1.8976	No
F2	> 95	13.5	0.7	0.071	2.3691	Yes
F3	> 95	3.1	0.6	0.111	3.7006	Yes
F4	> 95	2.7	1.4	0.277	9.2452	No

Note: Here, R_d, C_u, D_{15} and D_{c35} are relative density, uniformity coefficient, particle size at 15% finer, and controlling constriction, respectively.

limiting D_{15} are obtained and reported in Table 10.7. Accordingly, the allowable filter band is plotted in Figure 10.10a and various filter gradations (F1–F4) are selected within this allowable region, as shown in Figure 10.10b. A summary of geometrical parameters, including particle and constriction sizes as well as their ratios, is tabulated in Table 10.7. Filter gradation F3 is selected.

10.10.3 Design example 10.2: Geometrical assessment of internal instability potential of sub-ballast filter

Determine the internal instability potential for the gradations selected in the previous design example 10.1:

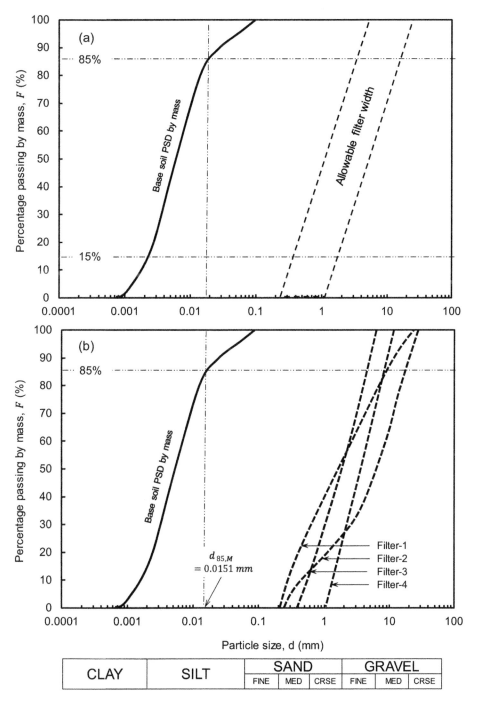

Figure 10.10 (a) Allowable filter band widths for embankment dam filter and railway sub-ballast filter layers for protecting the given base soil, (b) selected filter gradations

SOLUTION

Table 10.8 Summary of calculations for internal stability assessments

Filter	(H/F)min	D^c_{c35}	$d^f_{85,\,SA}$	$D^c_{c35} / d^f_{85,\,SA}$	Internal stability
F1	1.02	0.499	0.48	1.041	U
F2	0.98	0.770	0.4	1.923	U
F3	1.33	0.643	0.8	0.804	S
F4	1.71	1.37	1.95	0.701	S

Given that cyclic loading induces a significant agitation of granular media that results in variations in the constriction sizes, the modified CP–CSD criterion is deemed appropriate for assessing the internal stability of filters F1–F4 under cyclic conditions. Accordingly, the CSD of the coarse fraction is determined at $R_d = 0\%$ to anticipate the disturbance of filter media due to agitation.

This analysis reveals that filters F1 and F2 are geometrically unstable, while F3 and F4 are stable (Table 10.8). It is also revealed that filters F3 and F4 are safe against the inception of internal instability. Based on the above analysis, the following conclusion is drawn:

- Filter F3 is the most suitable option for a sub-ballast filter under given cyclic loading conditions.

Chapter 11

Appendix A
Introduction of SMART tool for track design

11.1 Introduction

SMART (supplementary methods of analysis for railway track) has been developed on the basis of knowledge acquired through two decades of laboratory studies, field observations and computational studies on rail tracks conducted at the Centre for Geomechanics and Railway Engineering (CGRE), University of Wollongong, under the leadership of the first author. The main user interface is illustrated in Figure 11.1. It contains research deliverables of numerous sponsored projects completed under the Australian Research Council, CRC, for Railway Engineering and CRC for Rail Innovation since the mid-1990s. Many of the concepts and analytical principles incorporated into SMART have been described by Indraratna *et al.* (2011b) and in this book in a design perspective.

SMART is a computer program written in MATLAB to aid in the analysis and design of rail track substructure. It consists of a comprehensive collection of performance-based methods for the analysis of track formation layers such as ballast, sub-ballast and subgrade. The user can select either individual or combined methods that are available in the software to perform routine design and analysis of ballasted tracks. SMART also allows the user to understand the effects of traffic characteristics and the properties of ballast, sub-ballast, subgrade and synthetic reinforcing elements on track performance. The program enables the user to determine the frequency of track maintenance based on the extent of deformation and drainage. SMART is a standalone computer program that contains user friendly interfaces for data entry and the presentation of results, as illustrated in Figure 11.2. The program can be installed and used on any 32-bit or 64-bit computers with Windows XP and later versions.

It is worthwhile to mention that SMART is not a replacement for existing design methods or for commonly adopted codes of practices. Where appropriate, it may be used as a value-added supplementary analysis in conjunction with primary design tools (e.g. numerical techniques), through research projects conducted at the University of Wollongong, Australia. SMART may be regarded as a valuable guide to assist the practitioner, but it should not be adopted as a complete design tool as its own.

If designers and practising engineers wish to purchase the SMART tool software, it is recommended to contact the Australasian Centre for Rail Innovation (ACRI) or the Commercialisation Unit of the University of Wollongong (George Tomka, gtomka@uow.edu.au) for more information.

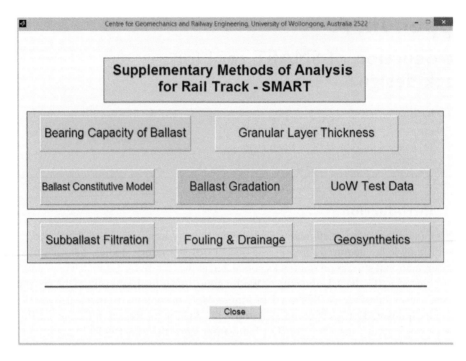

Figure 11.1 Main program window of SMART Tool containing eight modules

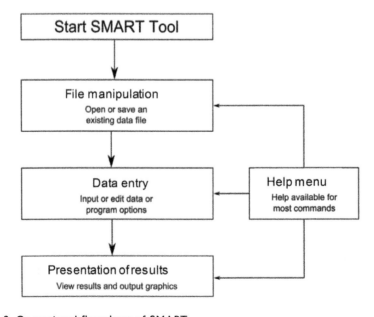

Figure 11.2 Operational flowchart of SMART

11.2 Practical design examples using SMART tool

11.2.1 Bearing capacity of ballast

The module to determine the "bearing capacity of ballast" is shown in Figure 11.3, where the allowable bearing capacity of ballast is shown along with the bearing capacity factors. This output window also shows the dynamic amplification factor and the static and dynamic stresses just beneath the sleeper.

11.2.2 Granular layer thickness

The SMART tool enables users to determine the required thickness of granular layer (ballast + capping) based on the Li and Selig method. After entering all the input parameters into the "granular layer thickness" window in SMART, click the "RUN" button at the bottom and the following output window will appear (Fig. 11.4).

11.2.3 Effect of confining pressure

After entering the values for the input boxes, the following output window appears when the user clicks the "RUN" button in the "UOW test data" menu. The plots between ballast breakage index (BBI) and confining pressure (σ_3) are shown in Figure 11.5.

11.2.4 Effect of ballast fouling on track drainage

Selecting the drainage module from SMART enables practising engineers to analyse the performance of ballasted tracks in terms of drainage. Figure 11.6 shows the typical input values for the permeability of the fouling material and clean ballast. When the user clicks the "RUN" button, the output values and plots will be displayed.

11.2.5 Effect of ballast fouling on operational train speed

To continue the analysis to assess the impact of fouling on the train speed, users can try the "fouling vs. speed" option built into SMART. Figure 11.7 shows the typical input parameters (for clay fouled ballast at 10 kPa; the confining pressure is from Table 6.4).

11.2.6 Use of geosynthetics in tracks

This module of SMART contains details of how to select suitable geogrids to reinforce the ballast, including the shear behaviour at the ballast–geosynthetic interface. It explains how the input parameters should be entered into the program, and how the output results can be used in the design of geosynthetically reinforced track. The geosynthetics module contains three sub-modules: "geogrid selection", "ballast deformation" and "geotextile filter". This module can be accessed by clicking the "geosynthetics" tab in the main window (Fig. 11.1), and then the following window will appear (Fig. 11.8 and Fig. 11.9). The user is required to enter the mean particle size of ballast (D_{50}) and the intended size aperture of the geogrid. To obtain the results, click "RUN" button.

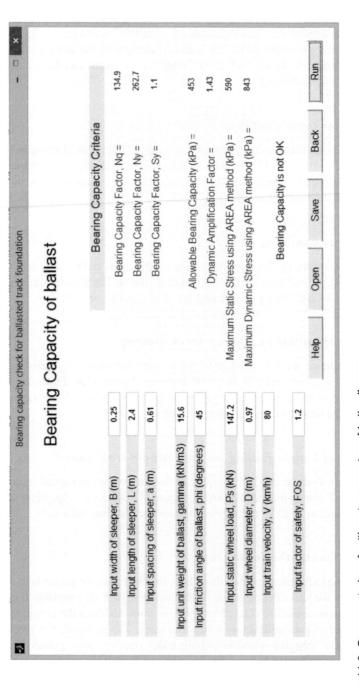

Figure 11.3 Output window for "bearing capacity of ballast" menu

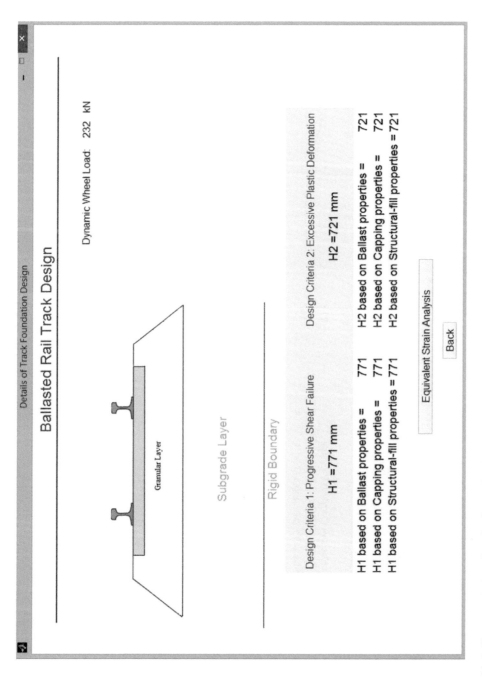

Figure 11.4 Output window for the "granular layer thickness" module

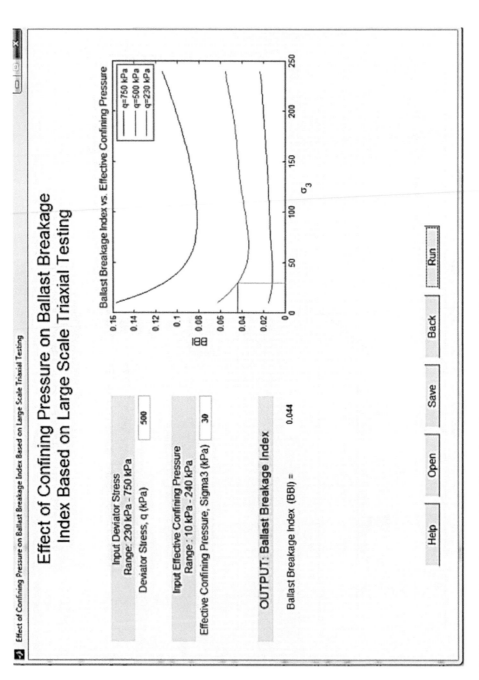

Figure 11.5 Output window showing the "effect of confining pressure on ballast breakage"

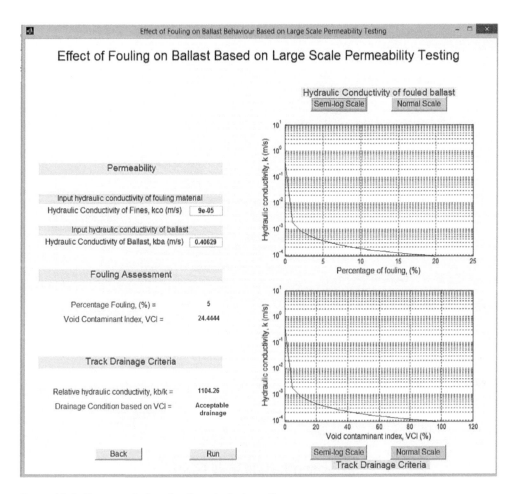

Figure 11.6 Output window for "track drainage"

11.2.7 Predicted settlements of ballast with or without geogrid

SMART can predict the deformation of ballast with or without geogrid. Figure 11.10 shows the input parameters used here for latite basalt (tested at UOW under controlled conditions with 300 mm thick ballast).

11.2.8 Ballast Constitutive Model

After entering all the parameters (13 parameters) into the "ballast constitutive model" menu, click the "RUN" button and the following output window (Fig. 11.11) will appear. This window shows the plots of deviator stress and volumetric variations with shear strain according to the input confining pressure.

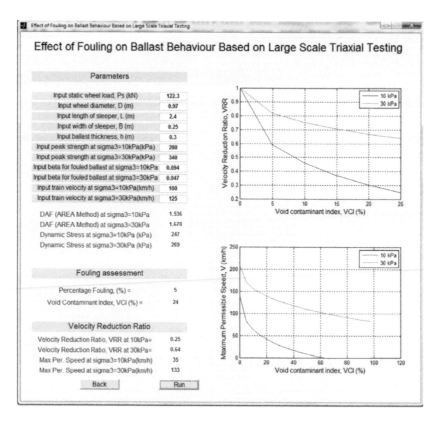

Figure 11.7 Output window for "fouling vs. speed"

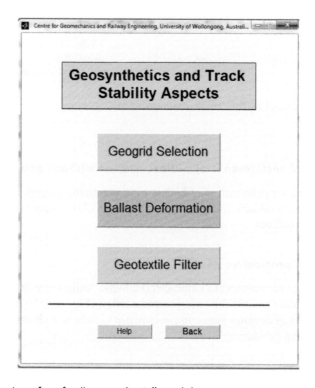

Figure 11.8 User interface for "geosynthetic" module

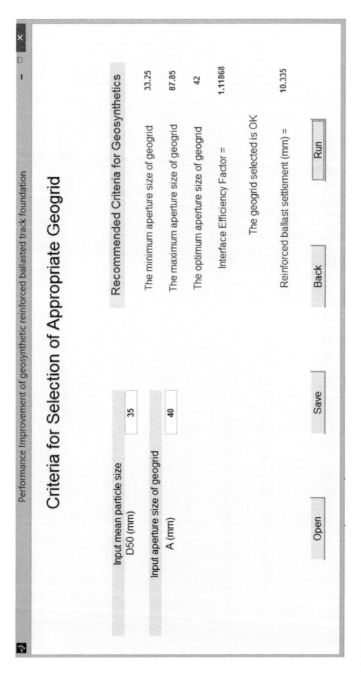

Figure 11.9 Output window for geogrid selection

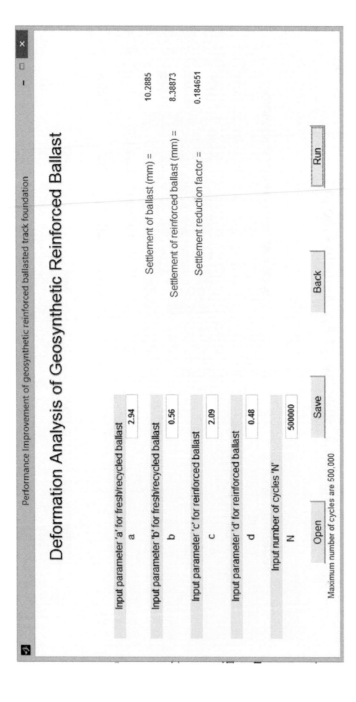

Figure 11.10 Output window for deformation of ballast

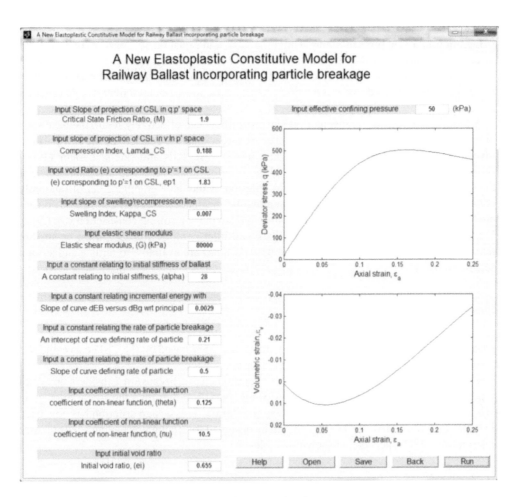

Figure 11.11 Output window for the "UOW test data" menu.

11.2.9 Selection of capping/sub-ballast for filtration layer

This module can be loaded by selecting "sub-ballast filtration" from the main menu (Fig. 11.1). Sub-ballast grading should be selected to prevent the subgrade material infiltrating into the ballast. It is noted that SMART adopts Terzaghi's filter design criteria for the analysis. The Terzaghi relationship (an empirical method) describing effective filters is used in SMART:

$$\frac{D_{15}}{d_{85}} \leq 4 \tag{11.1}$$

$$\frac{D_{15}}{d_{15}} \geq 4 \tag{11.2}$$

where the upper-case D denotes the filter particle size and the lower-case d is the base particle size (subgrade), and the subscripts refer to the percentage of particles that are finer than the size.

Equation 11.1 is to prevent piping within the filter, and Equation 11.2 ensures that the filter layer has a permeability that is several times higher than the subgrade.

Considering the size of the voids within the filter rather than the actual particle size is more appropriate when designing the filter, so a constriction size-based filter design criterion is imperative.

A filtration process using mechanical sieves as filters can effectively retain the base soils only if at least 15% of the base soil particles are larger than the aperture of the sieve. These investigations were mostly carried out on uniform base soils and filters. Lafleur (1984) showed that self-filtration takes longer with non-uniform base soils and more soil is lost if the filter is designed based on the Terzaghi criterion. Although a granular filter of randomly compacted particles is more complex than a regular mechanical sieve, it can still be considered to be equivalent to a sieve with apertures equal to the controlling constriction size (D_{c35}). From this point of view, for an effective base soil-filter combination, D_{c35} must be smaller than d^*_{85} to ensure that at least 15% of the base particles are available to initiate and sustain self-filtration, hence:

$$\frac{D_{c35}}{d^*_{85}} < 1 \tag{11.3}$$

where D_{c35} is the constriction size which is finer than 35% based on the CSD curve and d^*_{85} is the base soil size based on a surface area finer that 85%.

The above constriction-based criterion is comprehensive because it considers arrays of fundamental parameters, including PSD, CSD, C_u and R_d rather than the single filter grain size of D_{15} and the base size d_{85} in the Terzaghi criterion. This design criterion has now been incorporated into SMART. The filter should be designed for a subgrade having a particle size distribution (PSD) such as that shown in Table 11.1.

Trial 1

Enter the PSD for the sub-ballast as shown in Table 11.2.

After entering the PSDs for the subgrade and sub-ballast (i.e. filter), the user should click either the "PSD" or "CSD" button to update the results. After entering the relative density of filter, i.e. $R_D = 0.7$, click on the "RUN" button to see the results, as shown in Figure 11.12.

Table 11.1 Particle size distribution of the subgrade

Sieve size, mm	Percentage passing, %
0.001	5
0.01	10
0.075	35
0.15	61
0.212	80
0.3	94
0.425	98
0.6	100

Table 11.2 Particle size distribution for the sub-
ballast (filter) to be designed (trial 1)

Sieve size, mm	Percentage passing, %
0.15	0
0.212	1
0.3	2
0.425	4
0.6	7
1.18	12
2.36	25
4.75	45
6	65
9.5	100

SMART determines the constriction size distribution (CSD) in accordance with Terzaghi criterion.

Step 1: Calculation of Terzaghi criterion (refer to Raut 2006):

From the PSD of filter $D_{15} = 1.452$ mm

From the PSD of subgrade $d_{85} = 0.243$ mm

From the PSD of subgrade $d_{15} = 0.023$ mm

Using Equations 11.1 and 11.2:

$D_{15} / d_{85} = 5.98 \left(D_{15} / d_{85} > 4 \right)$. This ratio must be less than 4

$D_{15} / d_{15} = 63.1 \left(D_{15} / d_{15} \geq 4 \right)$

Therefore, Terzaghi's filter design criterion is not satisfied.

Step 2: Calculation of constriction size distribution filter design criterion:

$D_{c35} = 0.092$

$d *_{85} = 0.243$

$D_{c35} / d*_{85} = 0.37$

$D_{c35} / d*_{85} = < 1$ Therefore, CSD filter design criterion ($D_{c35}/d*_{85} \leq 1$) is satisfied.

The filter is not satisfied for Terzaghi's filter design criterion, but it is satisfied for the CSD filter design criterion. Users may have to re-design the filter.

Trial 2

The PSD of filter (sub-ballast) is changed, as shown in Table 11.3.

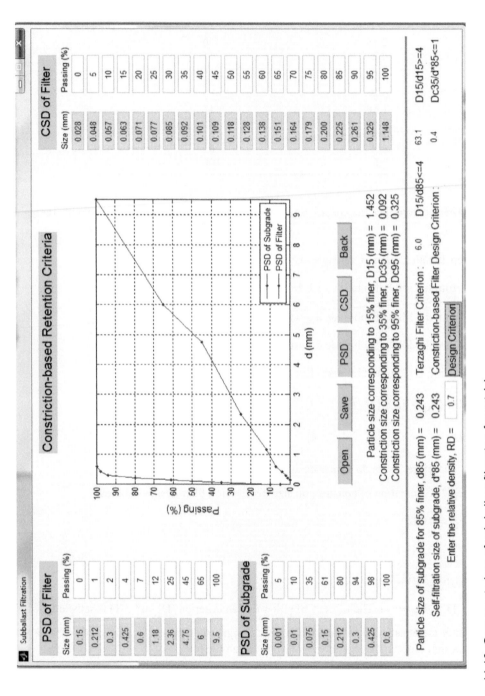

Figure 11.12 Output window of sub-ballast filtration for trial 1

Table 11.3 Particle size distribution for the sub-
ballast (filter) to be designed (trial 2)

Sieve size, mm	Percentage passing, %
0.15	3
0.212	10
0.3	17
0.425	24
0.6	33
1.18	54
2.36	73
4.75	85
6.0	93
9.5	100

After entering the PSDs for the subgrade and sub-ballast (i.e. filter), the user should click on either the "PSD" or "CSD" button to update the results, and then click the "RUN" button to see the results, as shown in Figure 11.13.

Step 1: Calculation of Terzaghi's criterion:

From the PSD of filter $D_{15} = 0.275$mm

From the PSD of subgrade $d_{85} = 0.243$mm

From the PSD of subgrade $d_{15} = 0.023$mm

Using Equation 6.76 a and b:

$$D_{15} / d_{85} = 1.13 \ \left(D_{15}/d_{85} < 4 \right)$$

$$D_{15} / d_{15} = 11.95 \ \left(D_{15}/d_{15} \geq 4 \right)$$

Therefore, Terzaghi's filter design criterion is satisfied.

Step 2: Calculation of constriction size distribution filter design criterion:

$$D_{c35} = 0.092$$

$$d *_{85} = 0.243$$

Using Equation 9.2:

$$D_{c35} / d *_{85} = 0.38$$

$$D_{c35} / d *_{85} < 1$$

Therefore, the CSD filter design criterion ($D_{c35}/d*_{85} \leq 1$) is satisfied.

The PSD selected for the filter satisfies the Terzaghi and CSD filter design criterion, and therefore the PSD for the filter (trial 2) can be used for the sub-ballast to prevent fine particles of subgrade moving into the ballast.

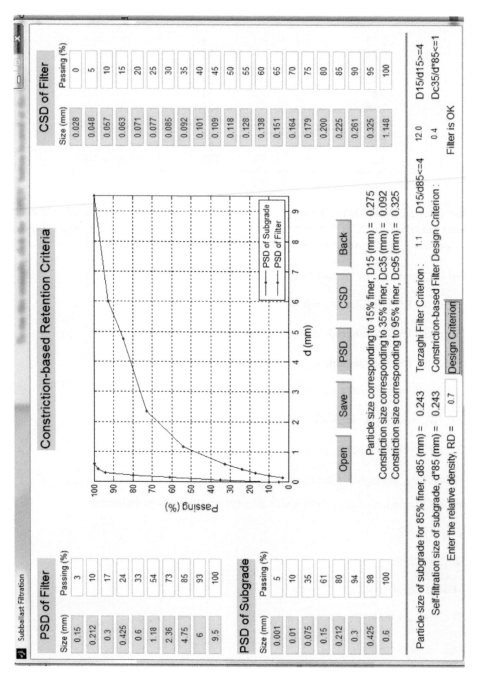

Figure 11.13 Output window of sub-ballast filtration for trial 2

Chapter 12

Appendix B

Unique geotechnical and rail testing equipment at the University of Wollongong

Construction of the National Facility for Cyclic Testing of High-speed Heavy Haul Rail at the Russell Vale High Bay Laboratory, University of Wollongong (funded by the Australian Research Council) – reinforced pit can accommodate sufficient depths of subgrade, structural fill or sub-ballast, and ballast for accurate simulation of dynamic track response.

Note: The super-strong frame can suspend two pairs of dynamic actuators (4-point synchronised loading) to impart the equivalent stresses generated by 35-tonne axle load trains travelling up to 200 km h^{-1} (i.e. very high speed for heavy haul trains that are often 4 km long in Australia).

Figure 12.1

Figure 12.2: installation of high capacity dynamic actuators of the National Facility for Cyclic Testing of High-speed Freight Rail; and Figure 12.3: view of 4-point dynamic actuator system.

Figure 12.2

Figure 12.3

Figure 12.4 National Facility for Cyclic Testing of High-speed Freight Rail

The iconic SMART – Rail laboratory

(results of numerous Australian Research Council (ARC) and CRC-Rail grants, and a $10 million research grant from the NSW Premier through RailCorp in 2009)

Large-scale Triaxial Testing System at University of Wollongong

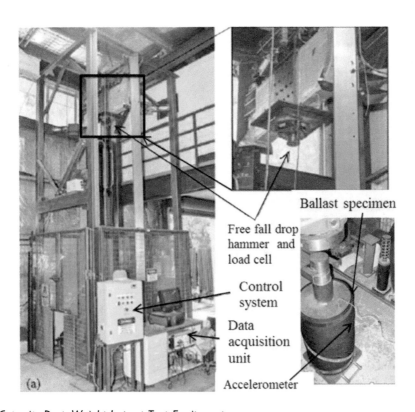

Ballast specimen

Free fall drop
hammer and
load cell

Control
system

Data
acquisition
unit

Accelerometer

(a)

High-Capacity Drop-Weight Impact Test Equipment

Unique Large-scale Track Process Simulation Apparatus

Large-scale Constant Normal Stiffness Shear Test Equipment

Large-scale Cyclic Direct Shear Test Apparatus

Large-scale Direct Shear Box (300 mm x 300 mm x 200 mm)

Large-scale Permeability Test Apparatus for Ballast Fouling

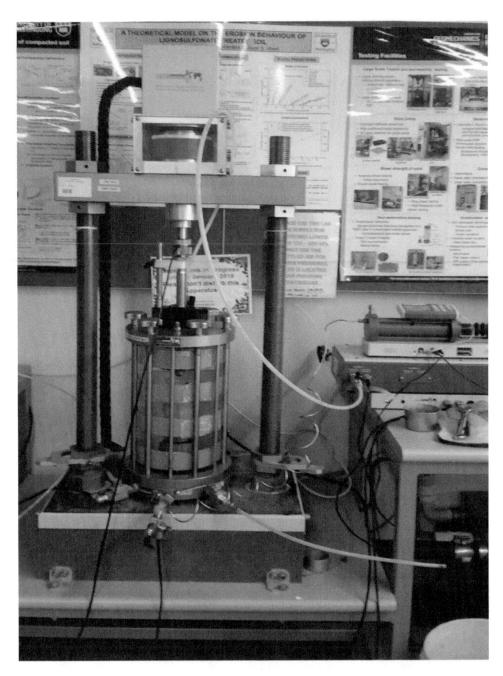

Large-scale Cyclic Test for Capping and Filtration Layer

References

AREA. (1974) *American Railway Engineering Association: Manual for Recommended Practice*, AREMA, Washington, DC.

AREMA. (2003) *Practical Guide to Railway Engineering, American Railway Engineering & Maintenance-of-Way Association*, Simmons-Boardman Publishing Corporation, Baltimore, MD.

Atalar, C., Das, B.M., Shin, E.C. & Kim, D.H. (2001) Settlement of geogrid-reinforced railroad led to failure due to cyclic load. *Proceedings of the 15th International Conference on Soil Mechanics and Geotechnical Engineering*, Istanbul.

Australian Standard: AS 2758.7. (1996) *Aggregates and Rock for Engineering Purposes; Part 7: Railway Ballast*, SAI Global, Sydney, NSW, Australia.

Bathurst, R.J. & Raymond, G.P. (1987) Geogrid reinforcement of ballasted track. *Transportation Research Record*, 1153, 8–14.

Bergado, D.T., Shivashankar, R., Alfaro, M.C., Chai, J.C. & Balasubramaniam, A.S. (1993) Interaction behaviour of steel grid reinforcements in a clayey sand. *Geotechnique*, 43(4), 589–603.

Biabani, M.M., Indraratna, B. & Ngo, N.T. (2016a) Modelling of geocell-reinforced sub-ballast subjected to cyclic loading. *Geotextiles and Geomembranes*, 44(4), 489–503.

Biabani, M.M., Ngo, N.T. & Indraratna, B. (2016b) Performance evaluation of railway sub-ballast stabilised with geocell based on pull-out testing. *Geotextiles and Geomembranes*, 44(4), 579–591.

Brown, S.F. (1974). Repeated load testing of a granular material. *Journal of Geotechnical and Geoenvironmental Engineering*, ASCE, 100(GT7), 825–841.

Brown, S.F., Kwan, J. & Thom, N.H. (2007) Identifying the key parameters that influence geogrid reinforcement of railway ballast. *Geotextiles and Geomembranes*, 25(6), 326–335.

Chapuis, R. P. (1992) Similarity of internal stability criteria for granular soils. *Canadian Geotechnical Journal*, 29(4), 711–713.

Chrimer, S. M. (1985) Considerations of factors affecting ballast performance. *Bulletin*, 704. *AREA – AAR Research and Test Department Report, No. WP-110*, 118–150.

Coleman, D.M. (1990) Use of geogrids in railroad track: A literature review and synopsis. *Miscellaneous Paper GL-90–94*, US Army Engineer District, Amaha.

Dombrow, W., Huang, H. & Tutumluer, E. (2009) Comparison of coal dust fouled railroad ballast behavior – granite vs. limestone. *Bearing Capacity of Roads, Railways and Airfields, Proceedings of the 8th International Conference (BCR2A'09)*, Taylor & Francis Group, London.

Ebersöhn, W. & Selig, E. T. (1994) Track modulus measurements on a heavy haul line. *Transportation Research Record*, 1470, 73–83.

Eisenmann, J. (1972) Germans gain better understanding of track structure. *Railway Gazette International*, 128(8), 305.

Esveld, C. (2001) *Modern Railway Track*, MRT Press, Zaltbommel, The Netherlands.

Feldman, F. & Nissen, D. (2002) Alternative testing method for the measurement of ballast fouling: Percentage void contamination. *Conference on Railway Engineering*, RTSA, University of Wollongong, Wollongong, Australia.

Ferreira, F.B. & Indraratna, B. (2017) Deformation and degradation response of railway ballast under impact loading – effect of artificial inclusions. *First International Conference on Rail Transportation*, Chengdu, China, July 2017, Paper ID: 362.

Fluet, J.E. (1986) Geosynthetics and North American railroads. *Geotextiles and Geomembranes*, 3(2–3), 201–218.

Han, J. & Bhandari, A. (2009) Evaluation of geogrid-reinforced pile-supported embankment under cyclic loading using discrete element method. In: *Advances in Ground Improvement: Research to Practice in the United States and China (GSP 188)*, ASCE, Virginia, United States

Hardin, B.O. (1985) Crushing of soil particles. *Journal of Geotechnical Engineering, ASCE*, 111(10), 1177–1192.

Hazen, A. (1911) Discussion of dams on sand foundation. *Transactions of American Society of Civil Engineering*, 73–199.

Hicks, R.G. (1970). *Factors Influencing the Resilient Properties of Granular Materials*. PhD Thesis, University of California.

Huang, H., Tutumluer, E. & Dombrow, W. (2009) Laboratory characterisation of fouled railroad ballast behavior. *Transportation Research Record: Journal of the Transportation Research Board*, 2117, 93–101.

ICOLD (International Commission on Large Dams). (1994) Embankment dams – filters and drains. *Bulletin*, 95, Nice: ICOLD.

Indraratna, B., Hussaini, S.K.K. & Vinod, J.S. (2012a) On the shear behaviour of ballast–geosynthetic interfaces. *Geotechnical Testing Journal*, 35(2), 1–8.

Indraratna, B., Ionescu, D. & Christie, D.C., R. (1997) Compression and degradation of railway ballast under one-dimensional loading. *Australian Geomechanics*, (12), 48–61.

Indraratna, B., Ionescu, D. & Christie, H. (1998) Shear behavior of railway ballast based on large-scale triaxial tests. *Journal of Geotechnical and Geoenvironmental Engineering*, 124(5), 439–449.

Indraratna, B., Israr, J. & Li, M. (2017a) Inception of geohydraulic failures in granular soils – an experimental and theoretical treatment, 227. *Geotechnique*, 68(3), 233–248 (doi:10.1680/jgeot.16).

Indraratna, B., Israr, J. & Rujikiatkamjorn, C. (2015a) Geometrical method for evaluating the internal instability of granular filters based on constriction size distribution. *Journal of Geotechnical and Geoenvironmental Engineering*, 141(10), 04015045 (doi:10.1061/(ASCE)GT.1943–5606.0001343).

Indraratna, B., Khabbaz, H., Salim, W. & Christie, D. (2006) Geotechnical properties of ballast and the role of geosynthetics in rail track stabilisation. *Journal of Ground Improvement*, 10(3), 91–102.

Indraratna, B., Lackenby, J. & Christie, D. (2005) Effect of confining pressure on the degradation of ballast under cyclic loading. *Geotechnique*, 55(4), 325–328.

Indraratna, B., Ngo, N.T. & Rujikiatkamjorn, C. (2011a) Behavior of geogrid-reinforced ballast under various levels of fouling. *Geotextiles and Geomembranes*, 29(3), 313–322.

Indraratna, B., Ngo, N.T. & Rujikiatkamjorn, C. (2013a) Deformation of coal fouled ballast stabilized with geogrid under cyclic load. *Journal of Geotechnical and Geoenvironmental Engineering*, 139(8), 1275–1289.

Indraratna, B., Ngo, N.T., Rujikiatkamjorn, C. & Vinod, J. (2014) Behaviour of fresh and fouled railway ballast subjected to direct shear testing – a discrete element simulation. *International Journal of Geomechanics, ASCE*, 14(1), 34–44.

Indraratna, B., Nguyen, V. T. & Rujikiatkamjorn, C. (2012b) Hydraulic conductivity of saturated granular soils determined using a constriction-based technique. *Canadian Geotechnical Journal*, 49(5), 607–613.

Indraratna, B., Nimbalkar, S., Christie, D., Rujikiatkamjorn, C. & Vinod, J.S. (2010) Field assessment of the performance of a ballasted rail track with and without geosynthetics. *Journal of Geotechnical and Geoenvironmental Engineering, ASCE*, 136(7), 907–917.

Indraratna, B., Nimbalkar, S.S., Ngo, N.T. & Neville, T. (2016) Performance improvement of rail track substructure using artificial inclusions – experimental and numerical studies. *Transportation Geotechnics*, 8, 69–85.

Indraratna, B., Raut, A. K. & Khabbaz, H. (2007) Constriction-based retention criterion for granular filter design. *Journal of Geotechnical and Geoenvironmental Engineering*, 133(3), 266–276.

Indraratna, B. & Salim, W. (2003) Deformation and degradation mechanics of recycled ballast stabilised with geosynthetics. *Soils and Foundations*, 43(4), 35–46.

Indraratna, B., Salim, W. & Rujikiatkamjorn, C. (2011b) *Advanced Rail Geotechnology – Ballasted Track*, CRC Press, Taylor & Francis Group, London.

Indraratna, B., Sun, Q.D., Ngo, N.T. & Rujikiatkamjorn, C. (2017b) Current research into ballasted rail tracks: model tests and their practical implications. *Australian Journal of Structural Engineering*, 1–17.

Indraratna, B., Sun, Q.D. & Nimbalkar, S. (2015b) Observed and predicted behaviour of rail ballast under monotonic loading capturing particle breakage. *Canadian Geotechnical Journal*, 52(1), 73–86.

Indraratna, B., Tennakoon, N., Nimbalkar, S. & Rujikiatkamjorn, C. (2013b) Behaviour of clay-fouled ballast under drained triaxial testing. *Géotechnique*, 63(5), 410–419.

Indraratna, B., Vinod, J.S. & Lackenby, J. (2009) Influence of particle breakage on the resilient modulus of railway ballast. *Geotechnique*, 59(7), 643–646.

Ionescu, D. (2004) *Evaluation of the Engineering Behaviour of Railway Ballast*. PhD thesis, University of Wollongong, Wollongong, Australia.

Israr, J. (2016) *Internal Instability of Granular Filters Under Cyclic Loading*. PhD thesis, University of Wollongong, Wollongong, Australia.

Israr, J. & Indraratna, B. (2017) Internal stability of granular filters under static and cyclic loading. *Journal of Geotechnical Geoenvironmental Engineering*, 04017012 (doi:10.1061/(ASCE)GT.1943–5606.0001661).

Israr, J., Indraratna, B. &Rujikiatkamjorn, C. (2016) Laboratory modelling of the seepage induced response of granular soils under static and cyclic conditions. *Geotechnical Testing Journal*, 39(5), 1–18 (doi:10.1520/GTJ20150288).

Jeffs, T. & Tew, G.P. (1991) *A Review of Track Design Procedure Volume 2 Sleepers and Ballast*, Railways Australia, Melbourne, Australia.

Jones, C.W. (1954) The permeability and settlement of laboratory specimens of sand and sand-gravel mixtures. *ASTM: Special Technical Publication*, 163, 68–78.

Kaewunruen, S. & Remennikov, A. M. (2010) Dynamic crack propagations in prestressed concrete sleepers in railway track systems subjected to severe impact loads. *ASCE Journal of Structural Engineering*, 136(6), 749–754.

Kenney, T.C. & Lau, D. (1985) Internal stability of granular filters. *Canadian Geotechnical Journal*, 22, 215–225.

Koerner, R.M. (1998) *Designing with Geosynthetics*, 4th ed., Prentice Hall, Upper Saddle River, NJ.

Koltermann, C.E. & Gorelick, S.M. (1995) Fractional packing model for hydraulic conductivity derived from sediment mixtures. *Water Resources Research*, 31(12), 3283–3297.

Konietzky, H., te Kamp, L. & Groeger, T. (2004) Use of DEM to model the interlocking effect of geogrids under static and cyclic loading. In: *Numerical Modeling in Micromechanics via Particle Methods*, Taylor & Francis Group, London.

Kwon, J. & Penman, J. (2009) The use of biaxial geogrids for enhancing the performance of sub-ballast and ballast layers-previous experience and research. In: *Bearing Capacity of Road, Railways and Airfields*, Taylor & Francis Group, London.

Lackenby, J. (2006) *Triaxial Behaviour of Ballast and the Role of Confining Pressure Under Cyclic Loading*. PhD Thesis, University of Wollongong, Wollongong, Australia.

Lackenby, J., Indraratna, B., McDowell, G.R. & Christie, D. (2007) Effect of confining pressure on ballast degradation and deformation under cyclic triaxial loading. *Geotechnique*, 57(6), 527–536.

Lade, P.V., Yamamuro, J.A. & Bopp, P.A. (1996) Significance of particle crushing in granular materials. *Journal of Geotechnical Engineering, ASCE*, 122(4), 309–316.

Lafleur, J. (1984) Filter testing of broadly graded cohesionless soils. *Canadian Geotechnical Journal*, 21(4), 634–643.

Le Pen, L.M. & Powrie, W. (2010) Contribution of base, crib, and shoulder ballast to the lateral sliding resistance of railway track: A geotechnical perspective. *Proceedings of IMechE*, 225, *Part F: Journal of Rail and Rapid Transit*, Special Issue Paper, 113–128.

Li, D. & Selig, E.T. (1998a) Method for railtrack foundation design. I. Development. *Journal of Geotechnical and Geoenvironmental Engineering, ASCE*, 124(4), 316.

Li, D. & Selig, E.T. (1998b) Method for railroad track foundation design. II: Applications. *Journal of Geotechnical and Geoenvironmental Engineering*, 124(4), 323–329.

Li, D., Sussmann, T.R. & Selig, E.T. (1996) Procedure for railway track granular layer thickness determination – Report No. R-898. *Association of American Railroads Transportation Technology*, 1–70.

Li, M. & Fannin, R.J. (2008) Comparison of two criteria for internal stability of granular soil. *Canadian Geotechnical Journal*, 45, 1303–1309.

Lobo-Guerrero, S. & Vallejo, L.E. (2005) Crushing a weak granular material: Experimental numerical analyses. *Geotechnique*, 55(3), 245–249.

Locke, M., Indraratna, B. & Adikari, G. (2001) Time-dependent particle transport through granular filters. *Journal of Geotechnical and Geoenvironmental Engineering*, 127(6), 521–529

Lu, M. & McDowell, G.R. (2006) Discrete element modelling of ballast abrasion. *Geotechnique*, 56(9), 651–655.

Marsal, R.J. (1967). Large scale testing of of rockfill materials. *Journal of the Soil Mechanics and Foundations Division, ASCE*, 93(2), 27–43.

Marsal, R.J. (1973) Mechanical properties of rockfill. In: *Embankment Dam Engineering*, Casagrande Volume, Wiley-Blackwell, New York, pp. 109–200.

McDowell, G.R., Harireche, O., Konietzky, H., Brown, S.F. & Thom, N.H. (2006) Discrete element modelling of geogrid-reinforced aggregates. *Proceedings of the ICE – Geotechnical Engineering*, 159(1), 35–48.

McDowell, G.R., Lim, W.L., Collop, A.C., Armitage, R. & Thom, N.H. (2008) Comparison of ballast index tests for railway trackbeds. *Geotechnical Engineering*, 157(3), 151–161.

McDowell, G.R. & Stickley, P. (2006) Performance of geogrid-reinforced ballast. *Ground Engineering*, 1(1), 26–33.

Nimbalkar, S. & Indraratna, B. (2016) Improved performance of ballasted rail track using geosynthetics and rubber shockmat. *Journal of Geotechnical and Geoenvironmental Engineering*, 142(8), 04016031.

Nimbalkar, S., Indraratna, B. & Rujikiatkamjorn, C. (2012) Performance improvement of railway ballast using shock mats and synthetic grids. In: *GeoCongress 2012: State of the Art and Practice in Geotechnical Engineering*, pp. 1622–1631, ASCE, Virginia, United States.

Ngo, N.T. (2012) *Performance of Geogrids Atbilised Fouled Ballast in Rail Tracks*. PhD thesis, University of Wollongong, Wollongong, Australia.

Ngo, N.T. & Indraratna, B. (2016) Improved performance of rail track substructure using synthetic inclusions: Experimental and numerical investigations. *International Journal of Geosynthetics and Ground Engineering*, 2(3), 1–16.

Ngo, N.T., Indraratna, B., Ferreira, F. & Rujikiatkamjorn, B. (in press) Improved performance of geosynthetics enhanced ballast: Laboratory and numerical studies. *Proceedings of the Institution of Civil Engineers – Ground Improvement*. DOI: 10.1680/jgrim.17.00051

Ngo, N.T., Indraratna, B. & Rujikiatkamjorn, C. (2014) DEM simulation of the behaviour of geogrid stabilised ballast fouled with coal. *Computers and Geotechnics*, 55, 224–231.

Ngo, N.T., Indraratna, B. & Rujikiatkamjorn, C. (2015) A study of the behaviour of fresh and coal fouled ballast reinforced by geogrid using the discrete element method. *Geomechanics from Micro to Macro – Proceedings of the TC105 ISSMGE International Symposium on Geomechanics from Micro to Macro, IS-Cambridge 2014*, CRC Press, Taylor & Francis Group, London.

Ngo, N.T., Indraratna, B. & Rujikiatkamjorn, C. (2017a) A study of the geogrid – sub-ballast interface via experimental evaluation and discrete element modelling. *Granular Matter*, 19(3), 54–70.

Ngo, N.T., Indraratna, B. & Rujikiatkamjorn, C. (2017b) Simulation ballasted track behavior: Numerical treatment and field application. *ASCE-International Journal of Geomechanics*, 17(6), 04016130.

Ngo, N.T., Indraratna, B., Rujikiatkamjorn, C. & Biabani, M. (2016) Experimental and discrete element modeling of geocell-stabilized sub-ballast subjected to cyclic loading. *Journal of Geotechnical Geoenvironmental Engineering*. 142(4), 04015100.

Nguyen, V.T., Rujikiatkamjorn, C. & Indraratna, B. (2013) Analytical solutions for filtration process based on the constriction size concept. *Journal of Geotechnical and Geoenvironmental Engineering*, *ASCE*, 139(7), 1049–1061.

NRCS (Natural Resources Conservation Service). (1994) *Soil Engineering*, National engineering handbook 26–633, U.S. Department of Agriculture, Washington, DC.

ORE. (1965). *Stresses in Rails*, Report D71/RP1/E, The Office of Research Experiment, Utrecht, The Netherlands.

ORE. (1969) *Stresses in the Concrete Sleepers*, Report D71/RP9/E, The Office of Research Experiment, Utrecht, The Netherlands.

Pender, M.J. (1978) A model for the behaviour of overconsolidated soil. *Geotechnique*, 28(1), 1–25.

Powrie, W., Yang, L.A. & Clayton, C.R.I. (2007) Stress changes in the ground below ballasted railway track during train passage. *Proceedings of the Institution of Mechanical Engineers: Part F: Journal of Rail and Rapid Transit*, 247–261.

Priest, J.A. & Powrie, W. (2009). Determination of dynamic track modulus from measurement of track velocity during train passage. *Journal of Geotechnical and Geoenvironmental Engineering*, 135(11), 1732–1740.

Qian, Y., Han, J., Pokharel, S.K. & Parsons, R.L. (2010) Experimental study on triaxial geogrid-reinforced bases over weak subgrade under cyclic loading. *GeoFlorida 2010: Advances in Analysis, Modeling & Design (Geotechnical Special Publication, 199)*, *ASCE*, 1208–1216.

RailCorp. (2010–2011) *Rail Corporation New South Wales, Annual Report 2010–2011*, Rail Corporation New South Wales, Sydney, Australia.

Railway Gazette. (2001) *BHP Breaks Its Own "Heaviest Train" Record*. Available from: www.railwaygazette.com/news/single-view/view/bhp-breaks-its-own-39heaviest-train39-record.html [accessed 23 November 2017].

Raut, A.K. (2006) *Mathematical Modeling of Granular Filters and Constriction-Based Filter Design Criteria*. PhD thesis, University of Wollongong, Wollongong, Australia.

Raut, A.K. & Indraratna, B. (2008) Further advancement in filtration criteria through constriction-based techniques. *Journal of Geotechnical and Geoenvironmental Engineering*, 134(6), 883–887.

Raymond, G.P. (1977) Railroad, wood tie design and behaviour. *Proceedings of ASCE, Transportation Engineering Journal*, 103, TE4, 521.

Raymond, G.P. (2002) Reinforced ballast behaviour subjected to repeated load. *Geotextiles and Geomembranes*, 20(1), 39–61.

Raymond, G.P. & Bathurst, R.J. (1994) Repeated-load response of aggregates in relation to track quality index. *Canadian Geotechnical Journal*, 31, 547–554.

Remennikov, A.M. & Kaewunruen, S. (2008) A review on loading conditions for railway track structures due to train and track vertical interaction. *Progress in Structural Engineering and Materials Incorporated in Structural Control and Health Monitoring*, 15, 207–234.

Rujikiatkamjorn, C., Indraratna, B., Ngo, N.T. & Coop, M. (2012) A laboratory study of railway ballast behaviour under various fouling degree. *The 5th Asian Regional Conference on Geosynthetics*, 507–514.

Rujikiatkamjorn, C., Ngo, N.T, Indraratna, B., Vinod, J.S. & Coop, M. (2013) Simulation of fresh and fouled ballast behavior using discrete element method. In: Indraratna, B., Rujikiatkamjorn, C. & Vinod, J.S. (eds) *Proceedings of the International Conference on Ground Improvement & Ground Control*, Research Publishing, Singapore, pp. 1585–1592.

Salim, W. (2004) *Deformation and Degradation Aspects of Ballast and Constitutive Modeling Under Cyclic Loading*. PhD thesis, University of Wollongong, Wollongong, Australia.

Selig, E.T. & Waters, J.M. (1994) *Track Geotechnology and Substructure Management*, Thomas Telford, London.

Shenton, M.J. (1984) Ballast deformation and track deformation. In: *Track Technology*, University of Nottingham, Nottingham, England.

Shukla, S.K. & Yin, J.H. (2006) *Fundamentals of Geosynthetic Engineering*, Taylor & Francis Group, London.

Sun, Q.D. (2015) *An Elasto-Plastic Strain-Based Approach for Analysing the Behaviour of Ballasted Rail Track*. PhD thesis, University of Wollongong, Wollongong, Australia.

Sun, Q.D., Indraratna, B. & Nimbalkar, S. (2014a) Effect of cyclic loading frequency on the permanent deformation and degradation of railway ballast. *Geotechnique*, 64(9), 746–751 (doi:10.1680/geot.14.T.015).

Sun, Q.D., Indraratna, B. & Nimbalkar, S. (2016) The deformation and degradation mechanisms of railway ballast under high frequency cyclic loading. *Journal of Geotechnical and Geoenvironmental Engineering, ASCE*, 142(1), 04012056, 1–12.

Sun, Y., Indraratna, B. & Nimbalkar, S. (2014b) Three-dimensional characterisation of particle size and shape for ballast. *Géotechnique Letters*, 4, 197–202.

Tang, X., Chehab, G.R. & Palomino, A. (2008) Evaluation of geogrids for stabilising weak pavement subgrade. *International Journal of Pavement Engineering*, 9(6), 413–429.

Tennakoon, N. (2012) *Geotechnical Study of Engineering Behaviour of Fouled Ballast*. PhD thesis, University of Wollongong, Wollongong, Australia.

Tennakoon, N., Indraratna, B., Rujikiatkamjorn, C., Nimbalkar, S. & Neville, T. (2012) The role of ballast-fouling characteristics on the drainage capacity of rail substructure. *Geotechnical Testing Journal*, 35(4), 1–11.

Terzaghi, K. & Peck, P.B. (1967) *Soil Mechanics in Engineering Practice*, John Willey & Sons, New York.

Timoshenko, S.P. & Goodier, J.N. (1970) *Theory of Elasticity*, McGraw-Hill, New York.

Trani, L.D.O. (2009) *Application of Constriction Size Based Filtration Criteria for Railway Subballast Under Cyclic Conditions*. PhD Thesis, University of Wollongong, Wollongong, Australia.

Trani, L.D.O. & Indraratna, B. (2010) Assessment of sub-ballast filtration under cyclic loading. *Journal of Geotechnical and Geoenvironmental Engineering*, 136(11), 1519–1528.

Tutumluer, E., Dombrow, W. & Huang, H. (2008) Laboratory characterization of coal dust fouled ballast behaviour. *AREMA 2008 Annual Conference & Exposition*, Salt Lake City, UT.

Tutumluer, E., Huang, H. & Bian, X. (2012) Geogrid-aggregate interlock mechanism investigated through aggregate imaging-based discrete element modeling approach. *International Journal of Geomechanics*, 12(4), 391–398.

Tutumluer, E., Huanng, H., Hashash, Y.M.A. & Ghaboussi, J. (2007) Discrete element modeling of railroad ballast settlement. *Proceedings of the 2007 AREMA Annual Conference*, University of Chicago Press, Chicago, IL.

Vesic, A.S. (1973) Analysis of ultimate loads of shallow foundations. *Journal of the Soil Mechanics and Foundation Division, ASCE*, 99(SM1), 45–73.

Walls, J.C. & Galbreath, L.L. (1987) Railroad ballast reinforcement subjected to repeated load. *Geosynthetics '87 Conference*. New Orleans.

Yang, L.A., Powrie, W. & Priest, J.A. (2009). Dynamic stress analysis of a ballasted railway track bed during train passage. *Journal of Geotechnical and Geoenvironmental Engineering*, 135(5), 680–689.

Index

Printed and bound by CPI Group (UK) Ltd, Croydon, CR0 4YY

22/10/2024

01777635-0007